Ups and Downs

Dear Jane,

As a nature-lover I hope this resonates with you - enjoy!

Best wishes,

Matthew

Ups
and
Downs

900 Kilometres on Foot
Through the French Pyrenees

Matthew Bowmer

Copyright © 2024 Matthew Bowmer

The moral right of the author has been asserted.

Apart from any fair dealing for the purposes of research or private study, or criticism or review, as permitted under the Copyright, Designs and Patents Act 1988, this publication may only be reproduced, stored or transmitted, in any form or by any means, with the prior permission in writing of the publishers, or in the case of reprographic reproduction in accordance with the terms of licences issued by the Copyright Licensing Agency. Enquiries concerning reproduction outside those terms should be sent to the publishers.

Troubador Publishing Ltd
Unit E2 Airfield Business Park
Harrison Road, Market Harborough
Leicestershire LE16 7UL
Tel: 0116 279 2299
Email: books@troubador.co.uk
Web: www.troubador.co.uk

ISBN 978-1-80514-211-9

British Library Cataloguing in Publication Data.
A catalogue record for this book is available from the British Library.

Printed and bound in Great Britain by 4edge Limited
Typeset in 12pt Futura by Troubador Publishing Ltd, Leicester, UK

Matador is an imprint of Troubador Publishing Ltd

Map of the Grande Randonnée 10

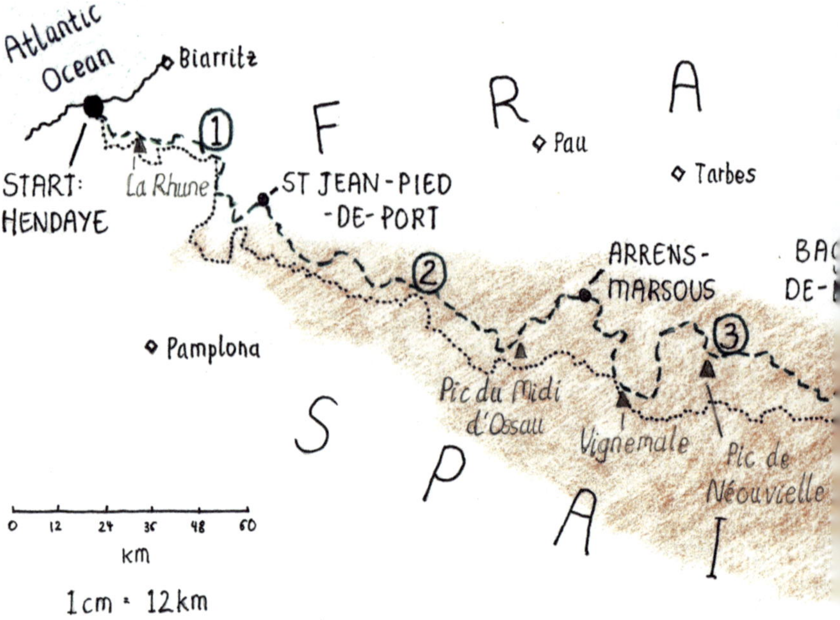
Atlantic Ocean
Biarritz
START: HENDAYE
La Rhune
1
F R A
Pau
Tarbes
ST JEAN-PIED -DE-PORT
2
ARRENS-MARSOUS
3
Pamplona
Pic du Midi d'Ossau
Vignemale
Pic de Néouvielle
S P A I
0 12 24 36 48 60
km
1cm = 12km

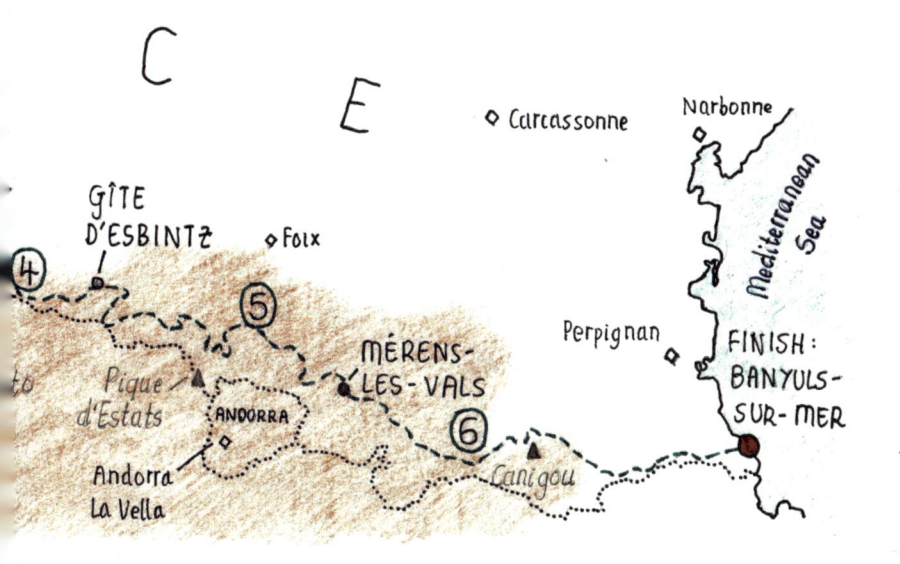

- - - GR 10 Route
- - - Country Border
① Section Number
● SECTION ENDPOINT
◇ Major Town /City
▲ Major Summit
High Altitude Area
Coastline

"I was free. I was affronted by freedom.
The day's silence said, Go where you will.
It's all yours."

Laurie Lee,
As I Walked Out One Midsummer Morning

Contents

Preamble   xi

1   Hendaye to Saint-Jean-Pied-de-Port   1

2   Saint-Jean-Pied-de-Port to Arrens-Marsous   24

3   Arrens-Marsous to Bagnères-de-Luchon   43

4   Bagnères-de-Luchon to Gîte d'Esbintz   60

5   Gîte d'Esbintz to Mérens-les-Vals   79

6   Mérens-les-Vals to Banyuls-sur-Mer   91

Postamble   119

Preamble

Preface

The Grande Randonnée 10 (GR10), also referred to as the *Sentier des Pyrénées*, traces a 900 kilometre route across south-western France, navigating the magical and unrelenting landscape of the Pyrenees from the Atlantic Ocean at Hendaye to Banyuls-sur-Mer on the Mediterranean. In doing so it passes through a variety of environments, from these coastal towns through temperate forests and rugged mountains which showcase the beauty of this little-visited range.

I can't remember exactly the first time I heard about the GR10, nor when I first decided I wanted to give it a go. It must have been shortly before the summer of 2013 because I can recall two distinct moments from that time. The first, wondering whether my GCSE French would be enough to manage the everyday situations and the more unique ones that were almost certain to arise. And the second, looking on Google Maps at the area around the coastal town of Banyuls at the end of the route, interested to see the landscape I might walk through on the final day. It seems strange now to picture the eagerness and positivity I pre-emptively associated with the vineyards and olive groves which dominate that section of the walk, because the reality was altogether a far more complex range of emotions than I could ever have anticipated before starting.

I did not attempt the path that summer and this initial seed lay dormant over many years until, for a few reasons that collided over the course of the preceding months, I found myself at Victoria Coach Station on the overcast morning of Monday 13th June waiting for the 8:30am BlaBlaCar coach to Paris. The following pages provide an account of the journey with its series of ups and downs.

In the text, * indicates the end of a day.

On My Back
- Eurohike 60 litre backpack
- North Face Tadpole 23 tent (2.3kg)
- Decathlon Forclaz 5°C sleeping bag
- Thermarest inflatable camping mattress
- Hi-Tech Ravine Lite walking boots
- Karrimor waterproof jacket
- Jumper
- Three old t-shirts for walking
- Hawaiian shirt (on the off chance I was invited to a village gathering and wanted to impress)
- Shorts
- Swimming trunks
- Four pairs of underwear
- Three pairs of Karrimor walking socks
- Bucket hat
- Sliders
- Stove

- Set of pans
- Gas canister
- Box of matches
- Small bottle of cooking oil
- Small bottle of washing up liquid
- Two 1.5 litre water bottles
- First Aid kit
- Roll of toilet paper
- Compeed blister plasters
- Hygiene bag (toothbrush, toothpaste, contact lenses and case, moisturiser, soap, sun cream, comb)
- Water purification tablets
- Towel
- Set of cutlery
- Five books including Paul Lucia's Cicerone GR10 guide
- A5 lined notepad
- Two pencils and a pen
- Route maps in a waterproof folder
- Compass
- Passport
- Headtorch
- Mobile phone and charger
- Wallet
- Earphones
- Camera and charger

Getting to the Start

A particular moist and affronting smell of collective body odour crossed with coffee. Panicked flapping of pigeons' wings. The scratching sound of wheels on the tiled floor. Raucous children drowned out by the collective din of a hundred conversations. Destination screens showing Brussels and Paris alongside Bradford and Plymouth. It could only be Victoria Coach Station.

My decision to take the coach down to the starting point at Hendaye was partly a result of my stingy nature and a sense of environmental guilt at flying, but there was also a significant part of me that was curious to find out what a 24-hour journey might be like. My advice, entertaining though the journey was, would be to avoid this course of action!

BlaBlaCar appears to operate a unique policy with their coaches, from Victoria Coach Station at least, where employing staff is an unnecessary expense and communicating with their customers is a privilege they are reluctant to provide. Having received an unhelpfully timed email at around 3am informing me the coach would be running 20 minutes late, I arrive at the station slightly sceptical about how smoothly this trip might run, especially given that I have a connection to catch in Paris. At 9am, half an hour after we are due to depart, with no further information besides an insightful announcement that our coach had not arrived yet, the busy waiting area at Bay 5 is beginning to fill with that peculiar British mixture of irritation but inaction. Tutting, eye-rolling, worried glances and even conversations with strangers. But apparently the idea of contacting BlaBlaCar to ask where our coach might be seems a step too far.

Sensing an opportunity to endear myself to fellow customers, I decide this inaction can go on no longer and call the company helpline, bracing for the confrontation ahead. None comes. Instead, I am faced with the classic problem of an endless string of answerphone options. "...

press 2 to amend your booking..." the voice whines at me, and I'm just giving up hope when I hear "...press 5 for an emergency involving one of our coaches". It appears this is the sole option for speaking to an actual human. The member of staff at the other end is surprised to hear that there's a delay and even more surprised to find he cannot locate the coach's current position because the GPS tracker has been switched off, although I appreciate his honesty in sharing this information with me. It amuses me to think that this coach has managed to break free from its traditional enslavement of running back and forth between European cities and could, at this very moment, be on any given trajectory, ready to carve out a new life in a foreign land.

Sure enough, though, the coach does arrive, one and a half hours late. At the wheel is the most relaxed coach driver I have ever come across. Somehow, through his laid-back demeanour, comical appearance and beaming smile, he seems to charm the whole cohort of passengers, myself included, into believing that the delay isn't actually a problem. It's almost as though by taking his time – completing the necessary safety checks in a slow and steady fashion, counting the number of passengers and pausing to ask the whole coach "*Alors on est prêt?*" – he convinces us that we aren't actually running late at all.

The result is that we set off with a remarkably positive level of energy on the coach. Better, it felt, than if we had left without mishap at 8:30am on the dot. Barry White's soothing voice fades in and out from the radio playing on the driver's dashboard, with occasional moments audible as the background hum of traffic momentarily eases off. My feelings quickly transition from the relief of having departed to the excited anticipation of the journey awaiting me, as I watch the view out the window becoming less and less familiar. Down towards the Thames, left onto Grosvenor Road along the river, past the towering Dolphin Square, over Vauxhall Bridge, curving round the Oval, through Camberwell and Peckham and soon enough we are shooting past a monotonous steam of semi-detached houses beside the dual carriageway that, in turn, gently become the green verges of Kent that hug the M20.

As the images on the other side of the glass become increasingly repetitive, my attention switches to the good-natured people who for one reason or another happen to be sharing this particular coach journey with me. Across from me sits a well-presented elderly man, on his way back to Paris after visiting his daughter in London, whose wiry white hair, leisurely movements and soft smile give him a saint-like aura. Ahead, just out of earshot, some young English climbers with brick-orange hard-hats hanging from their backpacks chat excitedly, the left earring of the guy nearest me bouncing now and then in animated moments. A few rows further back two strangers, both visiting London as tourists, strike up conversation about their respective experiences. I wonder what others saw when they looked at me; what my clothes, backpack, behaviour and expression might conjure up.

We pause at a service station just outside the M25. Perhaps the driver is tired after a mammoth half an hour at the wheel. In the stillness of the moment, bored by the toing-and-froing of people and vehicles in the car park outside and impatient to get going, I become conscious of the weight of my phone. Not so much a physical weight, but more its role in my life. Its impact upon me and a desire to break free from it. Now seems like my best opportunity to do so. I respond to a few messages, before putting it on flight mode in the hope I might become more present. It had occurred to me to adopt this policy for the duration of the walk, but this was one of those over-enthusiastic ideas that are all too easy to envisage before one has set off on the journey, and I would only manage a few days of abstinence before its weight exerted a pull on me once more.

Twenty minutes later we're ready to continue and the driver conducts his own version of checking all passengers have returned by inquiring "*Tout le monde est là?*". Happy to accept the resulting silence as confirmation that we were, he sets off with the side door still open, closing it as we enter the slipway to re-join the motorway. My mind is cast back to a friend whose son was left stranded at Watford gap services after getting off the coach for a cigarette. Before long we are running parallel to the Eurostar tracks. I look across wishing that affordable train fares between England and France existed, and shortly afterwards we reach Customs – a process which, for the coach, lacks the level of rigour I have experienced airports. Not only is there a breakdown in communication between the French coach driver, who speaks no English, and the English customs official, who speaks no French, but it turns out it is shockingly easy to bypass the entire passport check. Rather than entering the customs building after we disembark, it's possible simply to stroll straight across to the point where we reboard the coach, 50 metres down the road, as three passengers inadvertently do. Presumably they think it's just an opportunity to get a bit of exercise in before entering the Eurotunnel.

A sky of scattered cumulus clouds welcomes me to France as we emerge from the Eurotunnel, along with the underwhelming flat agricultural landscape that dominates the northern plains. In the end the changeover in Paris is comfortable enough which, for me at least, only serves to vindicate the coach driver's laid-back approach to a serious delay. Though no doubt my feelings would be different had I missed the connection. Briefly leaving the dingy and intense Bercy bus station for some fresh air, I pass an outdoor gym with rap blaring from a retro radio and a melange of muscular torsos, and saunter over to the *Passerelle Simone de Beauvoir* where the setting sun is shimmering brilliantly off the Seine.

From Paris I catch the San Sebastian-bound coach. Much to my disappointment it's not only packed, so that I end up being boxed in against the window seat, but there are no adjustments that might enable a relaxing night's sleep. Not that I was expecting pillows or bunkbeds, but optimistically I had envisaged it might have more legroom or thicker curtains. In my dreams. Well, hypothetically in my dreams, because it's impossible to fall into a sleep deep enough to dream.

As we set off from Bercy we head clockwise along the *périphérique*, passing a few places etched in my memory. Tower blocks near the commune of Ivry about which I had done a presentation while at university, and the setting for the beautiful film *Gagarine* by Fanny Liatard. The distinctive incinerator chimney at the Syctom waste treatment centre. Past a busy junction from which I started an optimistic hitchhiking trip to Switzerland a few years ago, which had been much more arduous and involved more stages than

I ever could have imagined. My enthusiasm for the views from the coach quickly wanes, as I contemplate the length of the drive ahead. The inability to move my legs or relax my neck to nap progresses from being mildly frustrating to downright painful.

Every two hours or so we approach one of our calling points and slow to a halt in the deserted streets. The harsh white lights come on and a muffled one-word announcement is made to inform us of the location: Massy, Tours, Poitiers, Angoulême, Bordeaux.... A few passengers alight and disappear into the network of streets, but many more take the opportunity for a cigarette break and even more still jump at the chance simply to stretch their legs for a few precious moments before we are beckoned back inside. The cycle continues in this way and I find myself glancing at every possible sign searching for the distance remaining to Bordeaux, which has the predictable effect of making the journey seem even slower.

*

Evidently my sign-searching proves boring enough an activity because my next conscious moment is waking up in the town of Dax, about 130 kilometres south of Bordeaux, shortly after sunrise on a cool, misty morning. Thankfully the adjacent passenger has got off so I am afforded the luxury of releasing my legs from a contorted trap, now able to sit awkwardly yet contentedly in a diagonal position across both seats. As we draw closer to my destination the scenery out the window seems more relevant, each new town and each new view feeling less like a meaningless transportation and more like the start of the adventure. The countryside is noticeably more verdant than I expected – all thanks to the thousands of kilometres of Atlantic Ocean lying uninterrupted to the west from which the oncoming clouds source their strength. As we continue on we pass the seaside settlements that have been built on its shores – Bayonne, Biarritz, Saint-Jean-de-Luz – by which point the foothills of the Pyrenees are coming into view. And finally to my destination: Hendaye.

The effect of a night on the coach is that by the time I alight I feel exhausted, but this is somewhat counteracted by my eagerness to get going. I stride along the seafront over to the starting point for the GR10 which, much to my confusion, is at the Casino Hendaye – not the French supermarket chain but a gambling house. I can't think of many places that are less compatible with the concept of walking hundreds of miles through unspoilt countryside. An elderly couple who have clearly sussed me out as a GR10-goer suggest that maybe the casino is an opportunity to earn a few thousand pounds to finance staying in comfortable accommodation for the whole walk. Tempting though it may be, I decide against taking the risk.

In my mind I was going to begin with a refreshing swim in the Atlantic, but it's not permitted this morning because of the large waves, and I'm not wildly keen on stopping to change into my trunks and sliders only to wade up to waist height before starting the walk. So, instead, I head to a nearby Carrefour for an unplanned and impulsive shop that I hope will last me until I next reach a shop, with most of my thoughts focused on getting out the town and up into the hills.

1

Hendaye
to
Saint-Jean-Pied-de-Port

98.5 kilometres and 4,680 metres ascent

Tuesday 14th – Sunday 19th June

Leaving the Coast

The path peels away from the coast, which I won't meet again until the Mediterranean 900 kilometres later, and ascends gently through the winding cul-de-sacs of the Hendaye suburbs. Overhead, the sun casts a series of stubby shadows from the adjacent fence posts that plot my steps into the countryside. I take a breather at the top of the first climb, at a dizzying altitude of 129 metres, feeling slightly out of breath and hoping that my stamina might improve in the days to come. It's a blurred mix of emotions – the thrill at having such independence and control over the expedition, the gratitude at having the time and money to be able to attempt the walk, apprehension about the uncertainty and distance ahead and mounting irritation at my increasingly sweaty back, all set amidst a backdrop of hazy fatigue following the drawn out journey.

    The Autoroute de la Côte Basque, the main route between Biarritz and Bilbao, carves intrusively yet with an air of elegance through the contours of the landscape. The thought of dodging six lanes of full throttle traffic with such a load on my back seems like a death wish, so I'm grateful the path passes beneath it via a subway. Its walls are adorned with amusing graffiti including, in capital letters for emphasis, 'I HATE SCHOOL'. From there it's not far to the enchanting village of Biriatou. Something about its perched hillside location, amphitheatre space and large conjoined houses all painted the same shades of red and white, give it the feeling of a smaller scale Tibetan fortress like Potala Palace, though I subsequently come to realise the architectural style is a common theme within the Basque country.

    Despite having hoped to walk further on my first day, drowsiness is getting the better of me and this seems like a logical place to stop. There's a water tap, public toilet and sheltered spot at the top of the amphitheatre-like area at the centre of the village, so I decide to spend the night here and hit the ground running tomorrow. It feels bizarre to saunter into the village and unpack my things so nonchalantly, but my newfound freedom on the path fills me with a cross between confidence and not giving a shit what anyone thinks. Only after spreading out my sleeping bag and devouring some brioche it dawns on me that the remaining six hours of daylight might drag with so little to keep me occupied – the options available being read, stroll around nearby or sleep. I opt for the latter.

Some time later I am disturbed from napping by a regular smacking sound and sit up to see two young guys playing a squash-like sport against the peach-coloured wall opposite. It brings the space in front to life creating a captivating scene which I feel an urge to film: panning round from the modest *mairie*, where I presume a meeting is occurring as a dozen or so people have arrived in quick succession in the previous minutes; to the scaffolding-clad *auberge* opposite where builders with a refreshingly laidback attitude to health and safety are adding the final few paint strokes before finishing for the day; to the stepped seating overlooking the central arena and the players hurrying around, shoes squeaking against the stone; to the azulejos antique shop, the young plane trees growing in a neat row by the car park and beyond to the generous gardens of the more modern houses at the edge of the village and a steep field where the farmer is gathering hay with his pitchfork. I can't help but think if it was in the UK it would be overrun with tourists. The scene and its tranquillity instil a deep sense of appreciation in me.

*

The billowing clouds and high humidity of the evening before led me to assume it would rain overnight, but I awake to an altogether different scene – mist in the valley and an intense sunlight that indicates it will be a hot day. I wash using the tap at the side of the village hall and start moving around 7am. Just as well because even at this time in the morning I am soon sweating heavily and getting irritable. The upside of this is that it prevents me going too quickly – a habit I find easy to fall into when walking alone, especially with a long distance ahead. Approaching the first summit, I am rewarded with a glimpse of the peaks that await me, stretching away eastwards towards the central Pyrenees, fading from the rich green of the bracken to a pale grey silhouette at the horizon. Any clouds have dissipated and the vast sky blue expanse stretches like a watercolour wash from east to west.

At Col d'Ibardin I'm surprised to come across a series of large supermarkets dotted among gîtes, bars and hotels, which mark the border with Spain, where shoppers from France take advantage of cheaper prices. Entering a two-storey shop complete with escalators and in a slight daze, I encounter for the first time the sensation of feeling out of place given my scruffy appearance, something that I hope will become more familiar and less embarrassing as the walk goes on. In the moment, though, I feel apologetic about my appearance and my perceived smell. Ridiculous, considering that I left Hendaye only 24 hours ago.

The remainder of the day features a pattern of regular breaks during which I remove my backpack and sit in the shade where my sweat cools with the aid of a gentle breeze. Then I trudge onwards, begin sweating again not long after and feel the need for another break. Each time I resist half-heartedly before succumbing and stopping once more. During one of these breaks, on a plateau beneath the distinctive La Rhune summit, two mountain bikers warn me that temperatures are only set to climb over the coming week. *"C'est le canicule"* they say with a resigned look and even their delivery sounds ominous.

Basque Pelota

On the gentle but boulder-strewn descent into the village of Sare is a stone trough formed from three magnificent slabs of moss-covered basalt. A small pipe overlain with rounded ceramic tiles feeds a humble trickle of water into the pool, that bubbles away musically to my thirsty ears. It couldn't be more appealing to a sweaty, overheating walker. My body feels as though it's already coated in a layer of dirt and I consider climbing inside and submerging myself for a few blissful moments of respite, but I'm put off by the voices from the garden a few metres away. So I settle for the compromise of dunking my head and soaking my t-shirt. At a spring less than a kilometre further along the path I repeat this ritual, before arriving in the village around 5pm as the light is fading from the glare of the middle of the day to the glow of the evening.

As I'm browsing the aisles of a Spar supermarket a gorgeous Mediterranean woman smiles in my direction. I'm transfixed. Too bad that she's merely a drawing on the side of a can of *edition limitée* Kronenbourg, commemorating the company's 75th birthday. In any case, I'm sold – apparently marketing does work. A stone's throw away, I find a good spot to rest and savour it, on the stepped seating of another outdoor theatre-like area. As I remove my boots and jot down a few thoughts about the day, I notice half-melted chocolate M&Ms dotted around at my feet from someone who must have been sitting here not long before.

The rectangular space in front of me features the same shaped wall as in Biriatou the evening before, except on a much larger scale and a court stretching back for almost a hundred metres with seating on either side. I watch, intrigued, as a man begins to drag a mat across the gravel surface, just as they do on the clay surface at Roland Garros in intervals between games. As he combs back and forth the evidence of his work is left behind in shadows that are so visible in the oblique evening light, reminiscent of pristine lawn where the grass has been mown in different directions creating geometrical forms. It's unclear whether this is in preparation for a game or if it's just an unusual daily practice for this man. My musing is answered in the next few minutes when two guys begin to play and, on this occasion, instead of using their hands or a racket, they have what look like hook-shaped gloves on one hand with which they catch the ball and hurl it against the wall in turn. Pretty soon quite a few other players have arrived, each wearing either maroon or green with immaculate white trousers. A small crowd gathers, and it transpires a game is about to commence.

The game, I find out, is referred to generally as pelota, this particular version being *grand chistera*, with the front wall called a *fronton*, a deep, narrow court measuring 16 by 80 metres, and all players wearing a wicker *xistera* on one hand. The warmup appears difficult enough, requiring quite an astonishing level of hand-eye coordination to catch the ball in the glove and launch it at the wall, but the game itself is on another level. The pace that's generated is such that it's often impossible to see the ball until it has rebounded off the wall and lost some of its speed, but equally impressive is the ability of the players to change the game suddenly by playing a 'drop shot' which bounces low at the front of the court, catching the opponents off guard. It's a special feeling to be learning the rules and tactics of a sport that's completely new to me, and all the more special to be doing so during a live game rather than being told about it or seeing it on YouTube. I even manage to get involved in the game myself when on several occasions a wayward shot sends the ball over the wall and flying into the village behind at which point, amusingly, play is interrupted while everyone, players and referee included, goes looking for it.

Almost as entertaining as the game itself is the interest passers-by seem to take in me. I guess I paint a picture of the quintessential traveller, relaxing there in the setting sun next to my boots and backpack, watching this spectacle attentively and looking down to scribble the occasional note in my diary. Or maybe I look like a down-and-out loner and they just feel sorry for me. First, a middle-aged couple pause right next to me and begin telling me about pelota, informing me that it's the fastest ball sport in the world, with the ball travelling at 200-300 kilometres per hour, and that there is a local league between the villages. Moments later I'm in conversation with a couple from Normandy who are on holiday here, whose children did Erasmus years in Southampton and Cardiff, and who are just as new to the game as I am.

After they continue on their evening stroll through the streets an elderly man passes by who, as soon as I start talking to him about walking, launches animatedly into a whole series of brief lectures, giving advice on everything from the route in the following days, to ensuring I stop to drink water often enough and even recommending a particular product which will help me replenish the salts lost during sweat. He shuffles away in a reluctant manner after this sweet but rather one-sided conversation, but when

I look over my shoulder a minute later he's back again, repeating largely the same advice especially about the danger that lies ahead on a certain ridge should I attempt to do it while feeling light-headed. I do my best to convince him I intend to take no such risks, while he fumbles around in his pocket and digs out an opened and half used packet of paracetamol as a parting gift. With that he wishes me good luck and disappears in the direction of home.

The willingness of strangers to talk and to listen is something that I am already finding so warming. They seem to do so in a manner that is patient to let you share your views, happy to give advice and the whole time managing to convey a genuine interest in and support of your journey. I feel a level of energy and appetite in this early stage which comes in stark contrast to the fatigue and alienation I experienced in London in the weeks and months leading up to my departure. Perhaps it's in part due to the enthusiasm we often channel during new chapters, but there has been an openness and approachability so far that is so welcome as a traveller on the path.

After the game I visit a bar in the centre of town in the hope of using the toilet and it transpires to be the happening place in Sare. The pelota players and two chunkier friends, who I struggle to picture skipping around the court, are tucking into pizzas in the corner. Inside, rock music is blaring from behind the bar where a scruffy bartender with a lumberjack-like appearance sombrely serves me *un demi* and it's all a bit at odds with the image of the traditional Basque village I have constructed in my head until now. I sit at a table outside, as far away from the speaker as I can get, and look across to the diners at the turquoise-shuttered

Hotel Arraya, sheltered beneath a dense row of pollarded plane trees. In the square beyond stands another, more neglected looking fronton. Over to the left a pavement sign advertises ice creams with small white stickers obscuring the prices of options I assume to be no longer available. There's an uplifting comfort while sitting here observing, as if purchasing the Pelforth fizzing appealingly on the table beside me has provided me access to this scene and allowed me to rest and reflect in a more fluid manner.

Post-poo, feeling lighter and more mobile, I go in search of somewhere to spend the night. A thick patch of lawn beside a wooden bus shelter attracts my attention which, despite being very visible, looks as though it would be comfortable enough. At first I try sleeping directly on my mat without the sleeping bag, and my jumper wrapped around my head as both pillow and eye mask, but it feels too exposed and in the end I pull the sleeping bag over me like a duvet even though the night is more than warm enough to do without it.

*

Le Canicule

I manage to get five hours of solid sleep, but at 3am I'm woken by the church bells and find that any areas of my sleeping bag that have been in contact with the grass have become wet with dew. Soon afterwards, a truck turns up and sets about restocking the Spar. Closing my eyes I can hear the repeated sequence of sounds – the tail lift of the lorry whirring as it descends, the forklift clunking over to the backroom storage and boxes being dropped into place –

followed by the driver trudging across the car park to use the public toilet and back again. When my alarm sounds shortly before 5am I'm not really in any mood to get up, so it's not until the 6am bells disturb me that I force myself to make a start.

I decide on such an early start because all the talk of hot weather has got me slightly worried. Setting off at this time has the added benefit that, for the first few hours, I have the path completely to myself as I pass through some stunning countryside that reminds me a lot of northern England. Pretty, scattered hamlets, neat and well-kept farms with extraordinary Basque names (Larretxekoborda, Itsasgaratekoborda, Sorroindokoborda, Etxegaraikoborda) on rolling land set against a backdrop of modest mountains in three directions. Around the halfway point the route runs alongside a river deep enough to bathe in and I need no second invitation. The sun has only recently risen and the water is bracingly cold, striking my flesh as I submerge, forcing me to draw breath and bringing goosebumps to ripple up my chest and arms. On the bed twists a network of roots which slip beneath my feet and conjure up a science-fiction goriness as though I might be swallowed into the underworld at any moment. Nevertheless, it's a beautifully restorative moment, cleansing my body seemingly both inside and out, and I feel a wave of cooling positivity as I set off wearing clean pairs of socks and boxers.

Ainhoa, where I arrive before 11am, is yet another beautiful Basque village featuring a pelota court at the centre and generous white houses with red shutters and exposed stone corners. Here, I decide it's time to check the weather forecast – something I have avoided until this point partly because there's an excitement in not knowing and partly because I feared any negative forecast would fester and become a source of unhelpful apprehension. Unfortunately, this proves to be the case when I see that temperatures will be rising to 40°C for the next three days, after which storms and wet weather are predicted to move in. As I expected, useful though this knowledge is, it sends my mind racing to consider every possible option available. For the first time I wonder whether I should just head home and return in friendlier conditions. Once I dismiss this option I entertain the idea of walking through the night to Bidarray, a settlement at the end of the next stage. Six sweaty hours pass while I mull my options over and doze on an uncomfortable dark green wooden-slatted bench in the welcome shade of a tree in the village square. Finally, I become fidgety and can't stand the indecision any longer, coming to the sensible decision to make my way gently up the next hill to a chapel and stay nearby.

There, as evening falls, I am reminded of the power that a short conversation can have to change one's mood. A number of campers are scattered around the chapel, one of whom is wandering through the small graveyard where the modest stones cast stretched shadows in my direction. I recognise her and her walking companion from earlier in the trip, though I can't pinpoint exactly where. On passing back to her tent we strike up conversation and it turns out she too is walking the GR10, along with her brother. There's a funny moment of uncertainty where I realise they're not French, and for a moment I think they might be English, which brings with it a combination of warm familiarity and cultural disappointment. But in fact they're German. Despite the differences in nationality, there's something magical about knowing they must be like-minded – why else would they have committed to this particular challenge – and about the peculiar odds that have brought us together to be in this place at exactly the same moment out of all the things you could be doing on earth. It's the first of many moments on the trip in which I feel a real connection with a fellow walker.

Sitting on the gently sloping ridge facing west towards the sun, still searing hot at 7pm, I can't help but feel slightly overwhelmed by a sense of spirituality – the natural beauty and religious energy combining to have a profound effect on me. In the distant haze I can just about make out the Bay of Biscay, my final glimpse of the sea before it becomes a blurred dream amidst a mirage of other images swimming through my mind. The peak of La Rhune, which I passed yesterday, takes centre stage. The railway to its summit is visible as a thin glimmering line snaking up its back. A small herd of horses descend past where I'm sitting to the copse around the chapel and, fearing the horses bells will keep me up at night, I move further up the ridge in search of solitude. There, for the first time, I pitch my tent. I consider preparing something using the camping stove, which would also be a first on this trip, but it seems ludicrous in such heat and so I opt for some baguette, a tomato and a tin of sardines for supper. Long before the sun sets, I crawl into the tent and wrap my improvised pillow and eye mask around my head, ready to shut off from the world.

*

Half an hour after closing my eyes a camper van rocked up with a group of noisy kids who cranked their speaker up and blasted any spirituality from the hillside. I built a mental picture of camp chairs and tinnies and began to believe any hopes of a good night's sleep had been dashed, yet early the next morning I wake feeling pleasantly refreshed and poke my head out of the tent to see no sign of them. Perhaps the spirits returned for vengeance.

Following the particular procedure of folding, arranging and packing my things into the backpack, I wave goodbye to my new German friends Hannah and Lennie, who I sadly never cross paths with again. The morning's walk consists of brilliant green ferns, sapping heat, ominously circling vultures and a treacherous descent down to a heavenly river which comes into view as an irresistible treat, sheltering a small community of walkers along its banks.

What plays out on many occasions during this stretch, most worryingly on the steep final descent to the river, is a curious battle between my desire to go quickly and finding a sensible speed that takes the heat into account. It's an early indicator of just how crucial your mindset is for long distance walking. I need to train myself to be patient despite my desire to reach the destination. What makes it so challenging is that once you overstep the mark and increase your pace it's very hard to slow down; firstly because it feels as if you would be prolonging your day by going any slower and secondly because making sensible decisions gets harder in the heat. The temptation to get to the village is getting the better of me and I am experiencing the early stages of heat stroke – struggling to place my feet with the precision necessary yet also refusing to stop before getting down to the river. By the time I reach it fifteen minutes later I'm desperate for water and shade.

Among the walkers at the river I meet Thomas and Solomon, two English guys around my age also attempting the GR10. At first there's something about them I find intimidating – perhaps it's their similarity to me. A bit like during Fresher's Week of university where a cohort of largely similar students size each other up and attempt to prove their individuality. I certainly didn't do this.... Solomon is tall and slim with a face that, despite its bristling unshaven nature, has an inherent sense of youth and seems ready to break out into a grin at any moment. He possesses an air of confidence, the sort of person who assumes a spokesman-like role in a group. Thomas comes across as shyer, more reserved, perching on a rock having just emerged from a pool in the river, showing his well-built physique and tattooed left arm. This strange feeling of assessing each other wears off as we chat and, during our exchange, an excitement slowly builds in me realising that every day I'm likely to encounter familiar faces from the previous few days, but never being able to predict who it would be or when it would happen.

The water is at first so inviting I'm tempted to spend the rest of the afternoon lounging here. But after my clothes are washed and then dried on a series of flat rocks seemingly designed for this very purpose (irritatingly losing a sock to the river in the process), I feel an inner restlessness and I'm compelled to continue. Wandering along in the dappled sunlight in no particular rush, I'm able to appreciate this stunning valley filled with towering beech, oak, sycamore and alder with the forceful torrent beneath rushing between calming pools. Here and there cars are parked along the road revealing tranquil spots hidden behind the dense vegetation, one of which to my surprise features a rotund woman paddling gently against the current on a monstrously large inflatable unicorn.

The path follows this course until it peels up to Bidarray, where I head in search of a bar with several intentions. Most pressingly to use the toilet, but halfway there and my bowels have other ideas, forcing me to dash into the woods for a quick pitstop, after which I experience such a strong sense of relief I imagine I can finish the whole walk that day if I just put my mind to it. I still need to charge my phone, though, which ran out of battery in the morning, having not been charged since I left London. I try to convince myself I'm enjoying living free from the constraints of time and, while being without it has been helping me feel more present, it has come with the unpleasant by-product of making me strangely disorientated and making simple decisions far more difficult than they should be.

I make a beeline for the church in the village, knowing I would find shelter there and I'm delighted to find a bar opposite. The cool interior with its thick sandstone walls and flagstones almost sucks me inside. As my eyes adjust to the dim light I realise the people at the table beside me are two guys I helped find the right path earlier in the day. At first I didn't recognise them because they're without their distinctive wide-brimmed sun hats. I'm trying to work out how they arrived here so soon when it dawns on me they could be the type of person who takes shortcuts and later claims to have done the whole path, and I feel a rush of irritation. 'You're not doing the full path, are you?' I want to blurt out, but it's a good thing I don't because they tell me they're here only for a few days, meandering along between villages. This I can accept. In fact, I have a pang of jealousy for the relaxed character of their holiday – no pressure to stick to the exact GR10 path, no worry about how your legs might hold up in the coming weeks, and the comfort of being able to splash out on gîtes and meals in between some leisurely walking.

It's a strange thing travelling alone. When you're travelling with a companion or in a group the opportunity to chat to strangers doesn't arise too often – either you're chatting with them or your presence with them creates an invisible barrier to outsiders. When travelling alone, because you're completely in control of your own choices and your own time, almost every meeting with a stranger can lead to an interaction. Simultaneously, from the perspective of others, you're more approachable thanks to your lack of company. So I've been finding that I can spark up a conversation with anyone and everyone. Perhaps I'm overthinking, but I've then found myself wondering whether I'm imposing on other people's time, people who have probably come on holiday to spend some quality time together and may not be too concerned with meeting others, especially an overly enthusiastic, smelly Brit. Here, for instance, as soon as I find out my shortcut-taking companions Matthew and Steve, are from the UK, I feel I could chat for hours on end. Equally, I don't want to be that person who never knows at what point to shut up. There's certainly an art in judging the right moment to start a conversation, but also in deciding when to wrap one up.

My efforts to shelter from the heat last until just after 6pm, when I'm too bored and impatient to remain. A quarter of an hour later, though, and already I'm building up an insufferable sweat, so I plonk myself down on a grassy section of path, plug my headphones in and wonder how much longer these conditions will go on. My plan is to spend tonight just below the summits ahead but the going has been painfully slow and I seriously doubt if I will get that far, even though it's only a few kilometres as the crow flies.

Soon after restarting I pass a beautifully sculpted rock, flat like a miniature altar, lying stubbornly right in the middle of the path, at which point I'm happy to follow even the most tenuous indicator that I should pause once more. It provides the perfect platform to set up the stove for my first cooked meal of the trip – tinned cassoulet and couscous. In the space of a few minutes a train of walkers, most of whom I recognise from bathing by the river, ascend past me at a speed which seems lightning fast compared to what I have been managing, all rather bemused by the meal being prepared. It's quite a palaver to get everything set up correctly, with spare clothes and kit tossed in a pile from my search for the stove, but once I'm eating any negativity has dissipated and the stodgy stew settles happily in my stomach.

I only make it a few hundred metres further up in altitude before coming across a solitary oak tree next to an outcrop of rock, with a little overhang that creates a cosy place for the night. I lay my clothes out to dry, stretched beside each other like some insensitive flag amidst the natural tones of nature. From my crevice I settle down to watch the sun setting over the ridge opposite.

*

Having been warned to face the hottest day of the trip, I set an alarm for 4:30am, hoping to make the most of the cool morning temperatures for the remaining ascent. It casts my mind back to two months earlier when I had worked briefly as a binman in Hammersmith and Fulham and had woken up at this time to catch the night bus through the deserted streets of central London. For some reason it's far easier to wake at this time in the Pyrenees.

It must be well over 25 degrees overnight and again it's suffocatingly hot inside the sleeping bag so I shift to lying on top of it. Yet this leaves me exposed, and I wake to find a slug halfway up my leg. It's horribly difficult to remove, clinging like a leech and leaving a slithery trail running up from my heel with no means of washing it off given my limited water supply and the scarcity of water on the hillside. Note to self: buy wet wipes!

I set off with a gentle pink sky behind me, peering over the foothills to the northeast, giving a false impression of softness, of comforting warmth. But all the while steadily intensifying until that split second when those first rays come stretching over and the relentless strength of yellow light takes hold, beats down upon the land, and all of a sudden you are reminded of the heat of the previous day, the stickiness, the relief of reaching water. How strange that I could crave this same sun on misty days!

In the town of St-Étienne-de-Baïgorry I lose another possession to a river. A slider on this occasion, which slips right off my foot as I'm bathing, floats swiftly downstream and, after a pathetic attempt on my part to chase it across the slippery riverbed, is swallowed up by the awaiting rapids, momentarily visible as it disappears into the distance. To console myself I build a harmonious image of my lost sock from the day before meeting with this slider at a confluence downstream and floating together out to the Atlantic Ocean. Fortunately there is an Intermarché around the corner where I pick up another pair. Although, while doing so, I attempt to joke about it with the cashier who seems to think I'm completely inept and asks, in a concerned tone of voice, whether I'm intending to walk all the way to the Mediterranean.

The afternoon, more than any other I have experienced in my life, brings the need to seek shade as the tarmac oozes, mirages fill the middle-distance, shutters close and thoughts become dizzy in the claustrophobic heat. Everyone and everything quietly surrenders in unanimous agreement. The flashing green cross at the pharmacy displays 45 degrees and I can believe it. I might have lingered longer in the town that evening, but the groups who passed me the evening before, including Thomas and Solomon and a trio testing out a long distance walking app called Hexatrek, are making a move and I am apprehensive about falling too far behind them. Even though so few days have passed I'm beginning to feel a reassurance from walking with others, and enjoying having the option to take a break from the silence that inevitably comes with spending so much time alone.

If I knew how fast they walked, however, I might have thought twice about joining them! Considering myself in good shape, I'm too proud to ask for a break so I keep quiet until one of them suggests doing so and I try to hide my absolute delight when they do. Only once we're stationary do the surroundings seep into me. The astonishing scale of the landscape, the deep incision carved by the torrent in the gulley far below us that has been relentlessly crashing away at the rock for hundreds of thousands maybe millions of years, the trees that have so successfully colonised these slopes and cling onto scree slopes and rocky crags at unthinkable angles, and the recognisable distant clanging of bells revealing the presence of the enduring Pyrenean cattle who possess an apparent indifference to the hostility of their home.

*

There's an immense comfort, that afternoon and the following morning with Thomas and Solomon, in being able to talk about familiar things, make cultural references and use colloquial language without searching for simpler synonyms. But most of all the ability to share a laugh which has an indescribable power after days of walking alone. Amidst the clouds, we come across a tent pitched on the flat ridge beneath the summit. Our voices must have been overheard because all of a sudden a head pops out, barely visible through the fog to screech "Hi, I'm from Southampton!", which is backed up by the statement "Sorry, I haven't seen any English people all week." Clearly I'm not alone in occasionally seeking out this familiarity.

Still chuckling from this unexpected interaction, we make the straightforward remaining ascent to the summit. There, a brief gap in the weather reveals a blanket of clouds stretching out to the horizon – a sensation that never fails to feel rewarding and fabulously detaching, as if all the problems of regular life have been isolated and left beneath. A ridge leads out tentatively ahead of us, only to be surrounded and gently swallowed by this wispy whiteness. In moments like this, the popular image of heaven existing in a separate space divided by a barrier of cloud appears self-evident, and there's a shared reluctance to descend back to the ordinary world.

Saint-Jean-Pied-de-Port

By this afternoon, six days since departing, the initial thrill of the trip is wearing off somewhat and fatigue is getting the better of me. So I succumb to a desire for comfort and book into a hostel in the centre of Saint-Jean-Pied-de-Port, saying goodbye to Thomas and Solomon who are continuing to the next village. The hostel manager, who is the only member of staff I encounter, must be somehow related to the driver from my coach ride down, because he's more intent on making guests laugh during check-in than worrying about the growing queue. It's a strange cross between a charming investment in welcoming each individual and the disrespect of keeping everyone waiting just so he could make sporadic comments like "carpe diem" and "remember to drink plenty of water". It may also have been an attempt at appeasement given the state of the dormitories – the one I'm allocated has no windows and an air conditioning system that, instead of gently humming in the background, saves up its noise for short explosive bursts every ten minutes. In any case I'm glad of a comfy mattress and my first shower of the trip which revives me.

The town is a popular tourist spot, not least because it marks a major converging point for walkers of the Camino de Santiago – a network of paths leading to the shrine of the apostle Saint James the Great in the cathedral at Santiago de Compostela. As a result, the centre is full of bars and I encounter the unfamiliar sensation of being spoiled for choice when I head out for a drink in the evening. I opt for one on the main high street with plenty of people-watching opportunities. I'm on the phone to my Dad when a man approaches and it's not until he's sitting at the adjacent table that I recognise him from the dorm. I overheard him earlier explaining to an American father and daughter how he'd quit his job in an environmental consulting firm because it was more about helping companies through loopholes in environmental regulation than avoiding their impact on the environment, so I already got the impression we would get along. When I look down to see he's walking around barefoot I only become more intrigued.

Unsurprisingly he's doing the Camino de Santiago. More surprisingly he's just broken a three day silence (incredibly also including avoiding any unnecessary noise such as from his walking sticks or phone), during which he has worn a sign hanging from his neck informing passers-by so as to avoid an awkward situation where they thought he was just being rude. He comes across with a level of enthusiasm and energy that screams of someone who has supressed an urge to speak for several days. But far from being nauseating, on this occasion it's a pleasure to witness. He tells me about spiritual techniques he employs on his journey such as taking cues from the environment to stop and breathe. Each time he sees a stream he pauses for a moment of appreciation. Maybe it's the effect of the beer, but in the moment I'm very caught up in the beauty of this idea, still very early on in the expedition and still possessing a largely patient and grateful mindset.

I'm finding our conversation deeply engaging but after a certain point a level of fatigue kicks in which coincides with one of his fellow Santiago walkers joining us at the table who I'm not wildly keen on. Feeling myself getting irritable, I leave them and take a short wander around the town, staring jealously at all the holidaymakers sitting down to three-course meals, only realising as I return to the hostel that I wandered off without paying for my beer. Any guilt is short-lived as I recall the frostiness of the barman – in fact it gives me a deep sense of satisfaction and I fall into a comfortable sleep back at the dormitory despite the regular rattling of the air con.

*

2

Saint-Jean-Pied-de-Port
to
Arrens-Marsous

176.6 kilometres and 9,950 metres ascent

Monday 20th – Sunday 26th June

Wild Beginnings

The mammoth day from Saint-Jean-Pied-de-Port to the cluster of buildings called Chalet Pedro feels like the boundary between two distinct types of landscape. The first on the whole tamed by humans with arable valleys and low summits that are reminiscent of many UK national parks I explored as a child. The second is rougher, more wild, dominated by the elements, difficult to access, with places where it seemed humans struggled to profit from the land. Testament to this are the many ruined hamlets, overgrown fields and orchards, young forests, makeshift fences and all-but-abandoned villages which have either turned to tourism to keep them afloat, been converted to holiday homes or become derelict.

The day is also split in two parts by the presence of a walking companion, Vincent, whose presence I quickly come to detest. I came across him in the Brasserie du Trinquet this morning where I stopped to satisfy a coffee craving which had been steadily rising since leaving Hendaye. The café had a cool dingy interior that would have appealed had I been arriving after a sweaty day's walk, but with my anticipation for the day ahead and the mild weather I chose to sit on the terrace and inspect the maps for the route awaiting me. There was something cinematic about the setting as Vincent came over to chat, cigarette in mouth and ashtray in hand. Somehow, following a few jumps in conversation he was telling me a long-winded story about going fishing during a thunderstorm which was delivered in a compelling manner but I only caught about half of the vocabulary. He also took a keen interest in the maps and asked to take a picture which suggested, bewilderingly, he would have set off map-less had it not been for this chance encounter.

Approaching midday I catch up with him in the quaint village of Caro, in conversation with a local. I offer a friendly greeting, hoping that I can glide on past without conveying any desire to continue together. Evidently this fails, though, as he responds with an enthusiastic "we meet again" and chooses to join me, at which point it also becomes clear that it would not be possible to walk in silence. It's not that he has an unpleasant character, but I exhausted a lot of my limited French chit chat at breakfast and I'd prefer to walk alone and immerse myself in the surroundings. So about an hour and a half later, when he makes an irritating niche reference to a cartoon cow that featured on French children's television I have obviously never fucking heard of, I decide I need to ditch him. We have lunch by the river in Estérençuby and I seize the opportunity to make a speedy exit as he's boiling water for a coffee. Wolfing down a hastily assembled baguette, I make my excuses and race away, hoping to put as much distance as possible between us to avoid the awkward possibility of meeting again.

Fuelled by this motivation, it proves to be one of the best days of walking of the whole trip. If only I could challenge feelings of frustration so effectively in everyday life! 19km later, with a thunderstorm developing behind me, I'm relieved to descend down to a road and my attention turns to finding a place for the night. There's a large, run-down wooden barn by the road with an air of the Bates motel from Psycho, made far worse by the unmoving silhouetted figures sitting at desks on the first floor and the thunder crashing overhead. My options for shelter are limited, though, so I manage, with a bit of effort, to force open a side-door to find a dingy, dirty but dry refuge space behind.

Clearly it has been well-used by walkers over quite some time, as it's equipped with several stained and holey mattresses divided across two 'bedrooms' separated by a low wall down the middle and a selection of plastic chairs around a fibreboard table into which names and dates of visitors have predictably been engraved. One corner is dominated by random objects piled up to the ceiling, presumably cast offs from local shepherds, in front of which stands a set of shelves with an array of bottles including an almost full orange liqueur called Suze. The other half of the ground floor is occupied by toilets and two showers, one of which produces only hot water and the other only cold. The wooden-slatted floor beneath the basins is overrun with sluggish toads and from the ceiling flickers the sort of white light one might associate with a sterile office environment.

An hour and a few glasses of Suze later I have largely forgotten about the creepy surroundings and I'm in a very jolly mood, having been joined by Thomas and Solomon, the girl from Southampton, Billie, and her boyfriend, Alex. We share an intangible sense of connection, able to exchange stories of what brought us here, our experiences so far and our anticipation of what the next few weeks might hold. Food wise, all that remains in my bag is a tin of cassoulet and all that remains in Billy and Alex's bags is a packet of rice. So between us we have a balanced meal which I enjoy, relishing the encounter and suspecting there will be few evenings ahead quite so sociable or cosy.

*

Overnight there's a bizarre situation with a kitten which must have been obscured behind the pile of discarded miscellanea and decides it wants to stand on my back as I try to doze off. Quite sweet in theory but in the moment I'm concerned it might somehow mistake my sleeping bag for a litter tray. Thankfully I awake to no such problem. The owner, one of the silhouetted figures I had seen on the first floor the evening before, comes down to see where it has got to and, given that I had written him off as a psychopath, is surprisingly friendly and easy to talk to.

The poor weather continues with a low cloud that hangs

stubbornly like a blanket, wrapped around the summits whose edges I skirt. There's nothing especially notable about the rest of the day barring a lost couple with one of the worst maps I've ever seen – showing only the outline of the paths and no other features – and a large group of drunken middle-aged guys having a barbecue packed onto a small platform, not to be put off by the conditions, one of whom shouts out to me "EET EES EEZEYER DOWN ZAN UP!". My thoughts turn to my shortage of food, entirely a result of my lack of planning up to this point, as well as the sensation of developing blisters, especially at my heels. These join the random mix of other things that churn around inside my head as I walk.

The hamlet marked 'Logibar' on the map turns out to be nothing more than a gîte d'étape and a car park. It's nevertheless very welcome after a monotonous descent on a difficult-to-follow path riddled with cow tracks. On arrival I consider rewarding myself with a half-pint but that seems far too stingy, so I'm delighted when the barman makes my mind up for me and automatically serves me *une pinte*. Annoyingly, the benches outside are plagued by flies which I use my remaining energy trying to swat before residing to my fate as they skip between my feet. (On countless occasions I have dreamed up elaborate contraptions I could invent for more effective killing of flies.) Yet in this context it would have taken a lot for me not to enjoy the beer. By the time the pint is finished any ideas I had of continuing have vanished and I book into a dormitory for a very reasonable €18, persuaded by the storms forecast and my recent wimpiness following two consecutive nights sleeping indoors.

The dorm is shared with two other guys meandering

their way eastwards according to their own schedules and preferences. Chris, an athletic German with a permanent apologetic expression completed the 4,270 kilometre Pacific Crest Trail a few years before, compared with which the GR10 seems like a stroll in the park. And Hendrick, originally from Poland but now living in Paris, who must be nearing 70 but professes to being new to long-distance walking. A refreshing example that it's never too late to start a new hobby, even one as demanding as this. He offers to buy me a beer claiming, tongue in cheek, that I saved his life earlier in the afternoon when I had done nothing more than point him in the right direction. Still, I accept the offer.

*

The forecast is proved accurate with a violent storm in the early hours, so I wake the next morning feeling that my choice to stay in comfort has been justified. Heavy rain persists so most of the guests hang around in the *séjour*. I help myself to some of the leftover coffee, bread and jam, while listening to a group discuss the rise of the Right and fractioning of the state and try to ignore the guy on the bench in front who is vigorously applying muscle soothing cream to his calves, only interrupting himself occasionally to offer a nod in support of one of the comments.

A break in the weather late in the morning acts as my invitation to set off, following a river on the banks of which I cooked my dinner of mackerel pasta the evening before, but which is now in full, deafening spate. Every now and then boulders or tree branches are distinguishable crashing along in the brown torrent. The forests appear to be breathing with a gentle yet constant vapour rising upwards to join the clouds, making it seem more like the tropics than the south of France, and I half expect to see monkeys swinging between the branches. At the Passerelle d'Holzarte, a pedestrian suspension bridge hanging 150 metres above the river with a clear view down through the chain-like structure, my fear of heights is really put to the test. It bounces playfully as I inch onto it and I need to pause to let my heart rate settle before fixing my eyes on the anchor at the other end and gently edging outwards. I proceed one step at a time, body horribly rigid, all other concerns eliminated, singing Andy Williams *Can't Take My Eyes Off You* in a desperate attempt to distract from the cavernous drop.

Besides this test, the day features an ongoing battle between the elements: heavy showers and ominous thunder rolling in, only to be dispersed half an hour later as rays of sunshine pierce through from above. My mood follows a similar but slightly delayed trajectory where the rain at first dampens my spirits but I gradually become accustomed to the conditions until they no longer carry the same miserable feeling. An improvement in the weather could occur that would bring with it a rapid improvement in mood. Yet, similarly, this too would wear off with time and the sunny conditions would soon be taken for granted.

Regardless of these fluctuations I am relieved to arrive in the group of settlements referred to collectively by locals as Sainte Engrâce. At its centre stands a miniature church consisting of a pitched roof tower and a main hall tapering down from it on one side. My intention was to camp nearby but, passing the gîte, I see Hendrick emerging from the doorway ushering me inside with such warm enthusiasm he might have worked here. In fact his welcome is far friendlier than that of the actual staff, who are incensed that I haven't booked (but not so incensed to turn me away) and reluctantly gesture to a small patch of grass that I could camp at the end of.

Maybe something in my expression of gratitude sounds insincere because just as I'm unclipping the tent a colleague approaches and brusquely instructs me to follow her to the dorm where there's an extra bed remaining, all the while muttering something about how there had been an increase in 'non-booking' customers. Strange etiquette given that I had merely asked politely if they had any space, and a request I thought they might have come to expect. It was almost as though paying guests were considered something of a nuisance. However, following some contemplation, I could see that dealing not just with tourists, but smelly and often impatient walkers day after day would be enough to push anyone over the edge now and then.

In the dorm I realise pretty much all the guests were at Logibar last night and I'm impressed how well everyone has walked to get here in such good time, until I hear that everyone except me has taken an easier route along the valley floor, apparently scared of the lightning. Once again I'm perplexed by their willingness to take shortcuts or alternatives in difficult moments. Fair enough a 100 metre detour to avoid boggy ground or find a good crossing point of a swollen stream, but to edit entire days only seems to cheat yourself of the satisfaction of completion. Why bother at all if a day of bad weather is going to put you off? Can you tell that this got on my nerves? Maybe I'm too uptight....

In any case I keep my opinions to myself as I'm clearly vastly outnumbered in my view.

Over a disappointing meal of tinned vegetables and cheap pasta, misleadingly called *cheveux d'ange* which couldn't be further from the truth, I spark up an entertaining conversation with a Dutch couple. The girl bears a spookily similar resemblance to a friend from university that I couldn't stop focussing on, to the extent that I become concerned she might think I'm looking at her strangely, which I probably am. They talk to me about the challenges of living in such a low-lying country including the cost of building raised roads, quoted as €35m for a 300 metre stretch on a recent project. The flood risk, so I'm told, has also produced a fairly widespread psychological fear of flooding among the population. I can't help but find this quite amusing but I do my best not to laugh as I imagine it wouldn't go down too well. When they turn in for the night I try to scribble down some further notes but the cool air of the evening makes the pages of my notebook damp, and my pencil no longer glides smoothly across the paper but instead gets caught up against it, so I too decide to call it a day.

*

A Jump in Altitude

With a big ascent lying in wait I had hoped to get a solid night's sleep but I end up adjacent to a chronic snorer, so I swap the noisy, comfortable dorm for the silent but rock-hard floor of the dining room and wake with rings under my eyes and back ache. I didn't make a breakfast order, nor would I have been allowed to, because breakfast is only for pre-booking customers. Of course. The only sustenance available is biscuits and chocolate, both out of date, but they could be useful to fuel the 1300 metre climb which begins almost as soon as I step out the front door.

The limestone landscape I pass through that day is nothing short of stunning and it seems appropriate not only that I'm listening to a podcast on the science of beauty as I march on upwards, but also that I have been reading Robert MacFarlane's *Mountains of the Mind*. In the book, MacFarlane writes about the Sublime, a term made famous by Edmund Burke in 1757, which refers to wild landscapes that go beyond beauty to a point that they become almost overwhelming. Certainly, it feels as if there has been a change in grandeur as I scramble up the Pas de l'Osque at almost 2000 metres above sea level and see the peak of Ahunamendi towering across the valley, patches of snow still clinging on obstinately despite last week's heat.

Nestled among rolling green hills and set against a backdrop of rocky peaks, made ominous by the clouds above, is the Refuge Jeandel, where my attention turns from taking pleasure in the scenery to the necessity of food. I scavenge some much-needed sustenance (bread, sausage and dried apricots) from their limited selection where the prices are anything but sublime. In the dining room two Australian guys tucking into hearty omelettes invite me to join them for a beer. I reluctantly turn down the offer as I can easily envisage all self-control rapidly evaporating and having to complete the remainder of the day's walk tipsy in the gathering darkness.

Shortly after reaching a bar in Lescun early in the evening, a heavy shower empties any remaining passers-by from the streets. There's something different about this place and I struggle to put my finger on it. I take my beer outside to sit beneath a parasol, immersed in the elements, and inhale that thick smell of rain I never tire of, all the while looking around in search of some obvious tangible explanation for my hunch. The architecture has undoubtedly changed along the way. The roofs here are steeply slanted and formed of slate tiles, the houses are less generously proportioned, the colours are more subdued, and it all combines to give a feeling of being the central Pyrenees, rather than the Basque country. At which point I realise this is the first settlement since I left the coast without a pelota court. While these courts might appear quite arbitrary and meaningless, they lend a sense of gravity to many of the villages, in much the same way that the tower or spire of a church does, and the obvious geographical influence of this sport feels somewhat significant.

In my notebook I start a sentence about how welcoming the décor is inside the bar, but when I return to order a second drink, I realise my views must have been exaggerated by my enthusiasm for the imminent beer, because the place is a peculiar hotchpotch of styles and objects. The furnishings are largely wooden with exposed beams, three curled fly catchers hang from one of these, littered with corpses. Affixed to a pale-yellow wall behind are two *chamois* heads above a ledge, on which a wide array of bottles stand, their number appearing doubled thanks to a mirror directly behind, and their labels largely unfamiliar to me. The main table is occupied by a group of scruffy yet friendly-faced English climbers doing the HRP (*Haute Randonnée Pyrénéenne* – an even higher altitude version of the GR10) whose good spirits fill me with a renewed strength to continue my walk, which has been faltering slightly over the course of the last 48 hours. On their way out they realise they need to stock up on supplies just as the épicerie opposite flips the sign and officially closes for the evening, and their polite English frustration ("damn, how annoying") is a source of cruel entertainment for me.

From the back room I can hear the barmaid's children screeching excitedly and it makes me wonder what it would be like to live here. Passing through these villages for an evening, it can be all too easy to think of them merely as a stage. A momentary pitstop on one's voyage, a place to sleep or stock up on food. In doing so, you can fall into a habit of romanticising their quaint appeal and get caught up in appreciating all the things that might be absent from one's own life – the peace and quiet, the community feel, the connection with the landscape, the relaxed pace of life – while ignoring the realities of actually living here, of spending every day waking up in Lescun. So I try to imagine the life of this woman, who must have been a similar age to me, working in this small bar and seeing visitors come and go all the time, living very differently to me, perhaps romanticising in the opposite way about a future life in a big city. I feel very lucky to get a glimpse into life in these places while having the freedom to continue on my journey.

Whenever I wrap my head up in my jumper it gives me that childlike feeling of security – if I can't see you, you can't see me. So it's fairly easy to settle on spending the night in the church porch, even though I feel a little uncomfortable when the caretaker comes to lock up at dusk and makes a distinct "*ahh non*" which is presumably directed at me tucked up in my sleeping bag in the corner. The upside to looking so down and out is that I'm much less likely to be disturbed or have my stuff nicked. Most people probably assume that anyone sleeping rough has nothing worth taking and, unless they wanted a 20-year-old camera, dirty walking clothes or random maps of the Pyrenees, they would be correct.

*

It's another night of poor sleep interrupted by the deafening church bells which, unlike in other villages, mark the hour and half hour throughout the night, as if locals might still be relying on the church to tell them it was 2:30am. After my morning routine of washing my face, brushing my teeth and packing my things, I still have some time to kill before the épicerie opens, so I sit on a bench outside the mairie doing a Sudoku on my phone, looking at the maps for the day, answering a few messages and filling my water. A school bus comes to a halt in the central square and a bike shoots down towards it, mum peddling wildly and child delicately balanced on the handlebars.

An ascent and half a descent later, the punishingly steep-sided valley above Borce provides an almost bird's eye view into the village, though it's a view that obscures any character, rendering it a dense form of grey edges vaguely arranged around a straight road. It feels somewhat like a satellite image of a linear settlement and, as I descend and look down every now and then, I have the impression I've got my finger on the scroll wheel of a mouse, steadily zooming in on the same image. At street level, being immersed among the buildings, it gains a sense of individuality. While it still possesses the same collective style as surrounding settlements – its small church, narrow alleys, lampposts, fountain, gardens and coloured shutters – you could now see the particular details that mark it out as unique. The evidence of personal taste and experience are clear in the assortment of geraniums, nasturtiums and marigolds springing from a planter, the distinctive umber shutters obscuring the windows of a house on the corner, and the homemade wooden sign advertising the 'vente d'œufs frais'.

From Borce to Etsaut it's a short stroll across a powerful river that runs north to eventually enter the sea at Biarritz. The place is alive with tourists, most of whom wear walking shoes of one kind or another, and it is a reminder of the community-like nature of the GR10. Not only do I bump into Chris at le Randonneur bar, who I'd first met in Logibar, but I'm also recognised by a bespectacled man named Michael who would be staying at the same gîte as me the following evening. There's a sort of magical intertwining of journeys, which weave in and out of each other, much of their stories similar but each very much a personal experience.

*

Into the Parc National des Pyrénées
A heavy overnight dew brings the unpleasant process of packing away a wet tent, which I have been hoping to avoid for as long as possible. Predictions of rain spur me on to leave early once again and maintain a good pace over the morning, past the Fort du Portalet and along a precarious

Ups and Downs

path above the menacingly named *Gorges d'Enfer* (Gorges of Hell) up into the Parc National des Pyrénées. Only when I reach the *col* at around 12:30pm do I feel able to relax in the knowledge that it's downhill from here onwards. Descending into the valleys during unpredictable weather always brings with it an unquestionable comfort, sometimes even a thrill that whatever the conditions throw at you, you will soon be sheltered in the woods or able to find a *cabane*.

In Gabas the abandoned Chalet des Pyrénées looms by the main road, a dense patch of weeds inhabiting its base. Only one of the three Es remain in the word Pyrénées, spelled out in a maroon typeface that looks as if it might have been designed by a primary school child equipped solely with WordArt. This has the effect of casting a rather miserable shadow over the cluster of buildings nestled together along this curving stretch of tarmac. Just down the road, at the most expensive épicerie on earth, where bitesize pieces of Emmental cost one euro, I enquire about rooms and the server happens to be the proprietor of the adjacent gîte. Once I've checked in, I am reminded of the luxury of simple things. In this order: a shower, a hob to cook with, a room to myself, being able to wash and dry my clothes and, a shame though it is to admit, WiFi.

*

I wake up early from a very deep sleep feeling slightly confused about the whole GR10 experience. There's a peculiar sensation that my whole life has been this walk and everything that came before was some elaborate dream, such is the level of immersion. It's as if my 'explorer' persona is taking over, my childlike desire to consider this as some sort of pioneering and all-encompassing quest steadily becoming reality. Or at least it certainly feels that way as I study the maps in the morning dawn, the dim yellow light from the corner lamp casting long shadows across the desk that fall off the edge and spill onto the floor. Rain patters lightly against the window and I'm reluctant to leave the warm snug surroundings.

There's a curious phenomenon that I encounter for the first time in Gabas, but that I would become accustomed to, whereby local people seem to be able to predict the weather more accurately than any forecast I could find online. I guess they are more familiar with the lie of the land and the very local level patterns of weather that modelling systems might not be so specific about. In any case, I find it not only very impressive but quite refreshing that this local knowledge is more reliable than internet sources. The forecasts for the day are very much full of doom and gloom, but the owner of the adjacent *auberge* believes the early drizzle will cease, the clouds will lift and it will be largely dry until around 3 in the afternoon. Sure enough it pans out this way, so that by midday I'm leaving the last wisps of cloud beneath me and shortly afterwards it seems absurd that this fluffy layer far below could ever have had an impact on my mood.

Up here at this altitude – approaching 2500m – the silence can take on an altogether different form. It's not the awkward silence that could fill a gap in conversation, nor the blissful silence of turning off a motorway into the countryside. It's somehow deeper, more permanent, more penetrating. At times it's welcome, heavenly even, and it fills your body with a sense of harmony and tranquillity. It can make you feel at one with the world, as though you wouldn't want to be anywhere but here. Yet, on other occasions it's a harsh silence, eerie, that can shock the body and make you desperate for any disruption – scree shifting beneath your feet, the high-pitched alarm call of a marmot, or the distant murmurs of human voices. Anything to interrupt the pervading, petrifying nothingness that you fear could continue forever and is always there lingering in the background. In either case there is an intensely beautiful quality to it.

In these moments of silence, one can achieve a meditative-like state in which thoughts come and go, seemingly out of your control. On this particular afternoon, the silence takes me back to Crouch End Arthouse where, two months earlier, I had watched a documentary called *The Velvet Queen* which follows French photographer Vincent Munier and writer Sylvain Tesson in a quest to film the elusive snow leopards of Tibet. The soundtrack by Nick Cave *We Are Not Alone* plays incessantly in my head and, in the emptiness of my surroundings, I wonder about the bears of the Pyrenees. As Tesson writes, "I was observed but unaware" and I can't help thinking that in the rocky slopes above a bear is looking down intrigued by this solitary figure trudging steadily onwards through its territory.

As the rain increases, clouds roll in and thunder starts, I pass swiftly into a forested area that soon leads to the ski resort village of Gourette – a strong contender for the ugliest place I have ever been. I'm hit by a sudden wave of late-afternoon exhaustion, but the only place I can find open amidst this jumble of concrete is a ski hotel, whose wooden interior bar shelters a group of four people playing cards who turn out to be the manager and his family. I order an espresso and attempt to make a sensible decision about what to do next, deciding to find a sheltered spot to settle down for the evening. Half an hour later I've left the hotel, settled down beside a bedraggled ping pong table and cooked a depressing meal of pasta with more pasta. This starchy overload combines with the shot of espresso to create a sudden burst of energy and convinces me to get out of this dump even if it means repacking my backpack and walking another 8 miles to the next village.

What follows is hard to describe. It's as if some higher power takes control of my body because I forget everything but the way ahead, skipping and jogging across the land, so that in just over two hours I'm descending towards Arrens-Marsous. Something that, in the moment seems to just happen, as if there was no question that I would reach my destination safely, with no thought for the risk I might be taking I hurtle down the narrow tracks, one slip away from injury in the driving rain.

Outside a sleepy bar I wring out my socks for the third time since leaving Gourette, suspicions confirmed that my boots' waterproofing has been compromised. I leave both socks and boots on a ledge by the entrance, not wanting to bring them inside dripping dejectedly behind me, and confident that no one will touch them with a bargepole. Inside, I order un demi much to the amusement of the fellow drinkers – two middle-aged big-bellied gourmand types deep in conversation at the bar who look as if they might have been here for some time.

It's a sizeable mental battle to leave the shelter of the bar and when I do, not long after, it doesn't feel like much of a victory. A few hundred metres away I find a dry patch of ground beneath the porch of the tourist office and settle down for the night, pulling on my jumper and jacket and regretting that I had only brought shorts with me. It's far colder than I could ever have imagined it would be at this time of year and it takes forever for me to drift off into a shallow sleep.

*

3

Arrens-Marsous
to
Bagnères-de-Luchon

183.4 kilometres and 10,253 metres ascent

Monday 27th June – Sunday 3rd July

Camping Dilemma

One theme that is beginning to occur too often for it to be coincidental, yet something I find difficult to admit, is that when fatigue begins to set in I am struggling to make rational decisions, which is only exacerbated by having no one else to consult. On some occasions I have made important decisions far too quickly and on others, I have deliberated for hours over the insignificant. This isn't to say that on the whole I've been disappointed by my progress so far. Far from it. But it has become acutely clear that some of my thought processes have gone slightly haywire without the stability of sleep patterns, diet, social interaction and other structures that make up what one might refer to as 'normal life'.

This morning is an excellent example. I wake up, pay a visit to the Proxi supermarket and impulsively buy all the ingredients needed for a sandwich, which I clumsily put together on a nearby bench and demolish in the space of a few gluttonous minutes. After the mammoth feat of the previous day, I'm feeling run down and look into the possibility of staying somewhere warm for the night. I phone the Maison Camelet twice without reply, research prices of three local B&Bs but decide they are out of budget, go to the campsite but don't feel it's right for me, walk to Maison Camelet to find it closed until the afternoon, and finally opt to sit in a café to get my thoughts straight. There, I manage to do some writing, respond to some messages and spend a while questioning my sexuality as I watch the rough-looking but well-groomed manager making coffees and pastries while conversing in French, Spanish and English. Dithering over, I opt to stay in the Pyrenees Nature campsite a few miles away. So I head back to Proxi for more supplies and then begin a steady trudge over the gentle pass to the next valley, by which time it's gone midday.

No sooner have I pitched my tent a few hours later than the drizzle sets in and it becomes apparent my tent – a sky blue North Face which must be around as old as me – is not sufficiently waterproof. It takes me a while to accept this fact, though, as I attempt to place plastic bags above the seams of the inner membrane and move my sleeping bag so it's no longer resting on the patches of groundsheet through which water is seeping. But with the prospect of heavier rain likely that evening I realise this is ludicrous. Packing up the tent only two hours after having erected it attracts the attention of a Dutch couple on the pitch next door. "Leaving already?". By this point I'm in a state of crazed hysteria, partly from exhaustion and partly the immediate uncertainty that lies ahead, which manifests itself in a kind of 'who the hell cares' attitude.

"What are you going to do now?" they ask reasonably.

"No idea!" I reply, mad grin etched across my face. My instinct is to find shelter in the hills, but it also crosses my mind to find the nearest bus stop and get the fuck out of here, thinking of all the welcoming comforts of home. Or even getting across to the end point at the Mediterranean, where I assume the weather will be drier, and working my way back to this point.

Thankfully I come to my senses and ask the reception whether they have a spare tent for the night, trying not to sound completely inept as I explain I have shown up in a tent that's not waterproof. As luck would have it they do have a spare, but I feel a little aggrieved to have to pay 10 euros, assuming I'll have to assemble this in the increasing drizzle. Luckily for me when I get to the pitch I see it's already been put up, and is like a palace – dry and spacious, featuring a porch with table and chairs, and blow up mattresses, set aside in its own quiet corner of the campsite. I recline with a beer and bag of peanuts marvelling at how my fortunes have changed in the space of half an hour, and grateful to be able to appreciate the evening colours beaming down on the pine trees across the valley.

*

Land of Glaciers

The tall, generous-scale buildings of Cauterets come as something of a shock when I reach them in the mid-afternoon on Tuesday 28th June. They offer extraordinary views down roads of Haussmann-esque architecture set against a backdrop of towering, forested hillsides which gives the impression I've been dropped into a toytown. It would hardly shock me if a shadow was suddenly thrown across the place as a gigantic figure stooped down, loosened one of the buildings from its current position with ease and relocated it elsewhere.

My efforts to find a bench to sit on and enjoy a drink by the river are sadly in vain as the torrent is channelled inaccessibly beneath the town centre and down the valley. In the end I resort to a spot nearby but when I make myself comfortable and crack open a beer I realise it's right outside a creche just as pick-up time is beginning. Damn it – I can't be bothered to move and I'm past caring about the funny looks cast my way.

Predictably the beer brings on a tired, happy and indecisive haze in which all concerns about the walk, and everything else in my life for that matter, feel wonderfully meaningless and I consider finding a quiet spot and dozing off. Inevitably, though, I eventually force myself to continue with the aim of getting to Cabane du Pinet – a shepherd's hut another 900m of ascent and 11km away. On the zig zag path out of the town, my backpack heavily laden with food, my momentum and optimism rapidly seep away so that after a few minutes of deliberation I decide the tent has to go. I have still been carrying it with the idea that it would provide a cosy barrier against the outdoors during dry conditions, even if it was useless in the rain, but it no longer seems worth the weight. In a few spontaneous moments I unclip the tent and carry it back down to the edge of the village where I prop it up against the side of a bin for a quick photo and final goodbye, before unceremoniously sliding it inside and, in the process, reminiscing about some of the places in which that humble structure has provided me with shelter and peace of mind.

Getting going once again, there's an immediate wave of relief at the weight loss and excitement at the prospect of having to fare without it. My mind is giddy thinking about the huge distances I will surely be able to cover now this encumbrance has been ditched. I plug into a podcast about breathing which offers a soothing respite from the monotony of my footsteps and my thoughts about reaching the destination. From the Pont d'Espagne the path climbs up into the Vallée de Gaube and what follows is possibly the most stunning landscape I have ever walked through. The route is cordoned off for maintenance but, concerned about taking a detour in the growing darkness, I slide past the barrier and charge on upwards past half-laid stones and unattended JCB mini diggers. As I ascend, the path's stone slabs are superseded by weathered masses of gneiss, and dramatic veins have been exposed on their face thanks to more resistant bands of minerals, pronounced by the acute angle of the evening light falling on its surface. These crisscross the path along with the roots of the surrounding pines, that together create a network of playful lines in the terrain guiding me onwards.

The blue of the Lac de Gaube in the twilight is unlike anything I have seen before and the entirety of the walk seems to be justified in a single moment. The glacial meltwater gives a milky light blue hue that looks almost artificial, as if food colouring has diffused thinly within it to leave a diluted cloudy finish. It lies harmoniously in this state, hardly a ripple on the surface, a blurred mirror image of the world above. Low hanging cloud sits in the valley above, creating a mystical light that is simultaneously dull and enchanting. From out of this cloud drops a thin yet striking cascade of white, carving silently and effortlessly down through its surroundings to touch the lake's surface, and then continues beneath in its reflection, into the depths.

I cannot face the prospect of having to make small talk that evening so I'm grateful to find the Cabane du Pinet mercifully empty. My only company is an energetic chaffinch that hops among the kindling at the door, chirping away excitedly as though we're long lost friends that have been reunited after years apart. With its perpetual strong smell of smoke, lack of light but for one tiny square window, literally rock-hard sleeping area and peculiarly located lock on the outside of the door, it's by no means luxurious, but I couldn't care less. I eat some bread and cheese outside, wondering what is happening at this altitude in other parts of the world.

*

My day gets off to a good start with a wild poo in a secluded spot. Not that I expect anyone to be passing so early in the morning, but it would be a horrible experience for everyone involved were I to be spotted in such an awkward crouching position. The route climbs steadily upwards, set against the dramatic backdrop of Vignemale, a towering series of rock faces of biblical proportions at the base of which stirs one of the few remaining Pyrenean glaciers. It rests there sullenly in the *cirque*, mostly sheltered from the brutal strength of the sun, but perhaps aware of the fate it is edging towards. It's a moment of beauty tinged with sadness in the knowledge that the bodies of ice here, which have had such a profound impact in shaping the landscape, have almost completely disappeared. From the mouth plunges a heavy mass of moraine spreading like a bedsheet with its folds and ripples to enter the smooth valley floor and reveal a surging flow of meltwater.

At the Refuge des Oulettes de Gaube a plaque commemorating the death of multiple climbers in an avalanche on the site in May 1982 reveals just how hostile these mountains can be. It plays on my mind as I traverse patches of snow on the ridge and subsequent descent, wondering how easy it would be for an accident to occur. At some point this summer this snow will become too weak to support a human's weight. Why couldn't that moment be now? And if it happens, then what? After a short while I thankfully manage to shake these thoughts off, acknowledging their futility while marching onwards unperturbed and even appreciating the resilience of these compacted snowflakes against the glaring sun.

A helicopter flies back and forth above, a pendulum-like basket swaying beneath it, transporting supplies up to one of the high-altitude refuges which cannot be accessed by vehicles. It's astonishing to think of the human ingenuity required to keep these places viable and the desire to seek out remote, inaccessible places. My destination for the evening is exactly that – remote and inaccessible – yet unfortunately without a cooked meal and a comfy bed waiting for me. The problem with these shepherd's huts is that of course they are without a booking system, so anyone can show up at any point. I've barely dried off from a bracing wash in a nearby stream, when I hear distant voices and moments later three young walkers and their dog appeared over the crest of the ridge above, resembling characters from a children's film with their flowing enthusiasm and excitable chatter. Presumably they're heading for the hut with the intention of staying the night here as well, so I stand at the door and wave from a distance, making my presence and message very transparent: no room at the inn!

Clouds are enveloping the peaks as the evening progresses and by time darkness falls a storm is brewing. The rain builds from a gentle patter to a fierce lashing that pounds down the chimney and onto the floor of the adjacent room and all the while the thunder, which at first can be heard far off, is edging ever closer until it's crashing overhead and seems to shake the thick stone walls of the hut, transforming them from a sturdy barrier against the elements to a flimsy membrane that could cave in at any minute. God knows how the three walkers are faring – my empathy does not stretch so far as to even consider going out looking for them. I close my eyes in a childlike fear and try to picture being anywhere else. Gradually the storm moves onwards, though the sound of distant thunder continues well into the night and keeps me in a state of underlying anxiety until the morning sun restores a sense of calm.

*

Up until this point I have been carrying five books in my bag, something which has shocked anyone who happened to see the contents of my backpack, given the weight-saving extents most people go to. My notebook, Paul Lucia's Cicerone guide to the GR10, *Mountains of the Mind* by Robert MacFarlane, a series of chapters by different authors about the experience of being an immigrant in the UK called *The Good Immigrant*, and *The Sun and Her Flowers* – a collection of poems by Rupi Kaur which I had bought with a book token several years back and never delved into.

Before arriving in Hendaye I had the wholesome idea of reading a poem each morning in the belief it might create a healthy, almost meditative routine, but two weeks into the trip and not having looked at the book once, it's time to address this. Over breakfast of an apple, some dates and three processed supermarket croissants I rather therapeutically select and carefully tear out all the poems which appeal to me and put them inside a waterproof folder, leaving the remainder of the book behind in an attempt to shed further weight. Despite my best efforts it stubbornly refuses to catch alight, so a scavenged, singed carcass remains for the next walker to find and wonder about.

In Luz-Saint-Saveur, eight hours and 32km of winding path later, I fall into a tourist trap, paying €4.40 for a half pint of Grimbergen Ambré in a bar with no plug points and no WiFi. I would be more pissed off but it tastes unbelievable. I continue to treat myself at the *auberge de jeunesse*, booking a *demi-pension* for the first time – bed, dinner and breakfast. The proprietor is without a shadow of a doubt one of the most enthusiastic yet indecipherable people I have ever met and possesses an ability to deliver every sentence with a level of gusto and personability that would exhaust any normal personal in the space of fifteen minutes. Maybe this is an attempt to make up for the ramshackle premises which feature, among other curiosities, a sizeable crack in the floor, ramps in unexpected places on the first floor and a fire door by the toilets that didn't fit the doorway, which allows a pleasant draft through but must prove a serious insulation problem in winter.

I optimistically think I've got the dormitory to myself but shortly before dinner a group of sweaty cyclists come crashing in making a staggering racket. Soon enough the room is full of their damp, smelly gear hanging off every possible hook in the room and they're grumbling to a member of staff, who looks like the hunchback of Notre Dame but without any hair, about the lack of towels. In the end we're all sitting together at dinner, a four course feast including a fabulously tender lamb stew and the overrated *gâteau Basque*, all delivered by the three-man team with remarkable flamboyance considering they must have to do this night after night. The cyclists are actually quite a friendly bunch, though they suffer from the widespread condition of not having much to say to a stranger, which in a lot of scenarios I have no problem with, but when seated around a table together at a meal ends up being slightly awkward. Towards the end of the evening the proprietor comes over to our table for a conversation which becomes progressively harder for me to follow at which point I make my excuses and retreat to the dorm.

*

Bovine Encounter

There should be a snoring test for anyone staying in a dormitory. After another disrupted sleep and a desire to eat my breakfast undisturbed, I'm happy to lie in bed and wait until the group has finished eating so that I can get some time to myself. When I do finally descend to the dining room I leaf through the surviving Rupi Kaur poems, finding solace in the simplicity of *Sunflowers*, all the while doing my best not to gulp the beautifully smooth double espresso that accompanies a small mountain of bread and jam.

Many tiring hours of ascent later I'm surprised and hugely disappointed to pass a building site up at 2200 metres altitude, but pleased to realise it looks to be for a new refuge. A man armed with a pickaxe is smashing up rocks in 30°C heat with no shade, and any internal complaints I have about being unpleasantly sweaty rapidly evaporate. At the cabane where I intend to spend the night I hear from a fellow walker that there are other people hoping to sleep there too. So it's a huge relief when I find out they are a sweet trio who were at the auberge the night before: an old man whose large, rounded nose and thick square glasses remind me of the protagonist, Carl, from the Disney film *Up*; and his two children. There is, in fact, what feels like a small community of walkers staying by the Lac de Coueyla-Gran, most of whom are dotted around in humble tents and behave in such a way that is the perfect middle ground between the extremes of in-your-face and standoffishness. Happy to chat yet also able to afford each person their own space.

After a dip in the lake, which is kept short as the nail on the big toe of my right foot threatens to tear off, I wash my clothes and spread them to dry on a smooth rock oriented towards the late afternoon sun. But as I'm checking on them an hour later I become surrounded by a herd of stealthily moving cows ascending up the valley. They force me to make a hasty exit, intimidated by their large horns and unsettling grunts, presumably conveying their disdain at my uninvited presence on their turf. My unguarded clothes promptly become the focus of their attention, and must have held a salty aroma which appeals to them, because the rock is soon encircled and by the time the cows move on my t-shirts are wet with saliva (but thankfully nothing else!). It proves to be a hilarious spectacle for the onlooking campers and once I reclaim my clothes I too can laugh about it, especially as there remains enough sunlight to re-wash and re-dry them.

My body clock has been slowly adjusting to the natural daylight as I've spent so little time indoors since leaving Hendaye. As dusk progresses I tend to feel fatigue catching up with me and, like many evenings on the trip, I head to bed at 9pm, before the sun has fully set, weary-eyed and prepared to forego the spectacular light show filling the sky.

*

Seeking Shelter
A steep and relentless descent leaves my knees in absolute agony on the approach to Vieille-Aure, a village which, when pronounced quickly, sounds like VAR – the acronym for Video Assistant Referee for anyone that doesn't follow football. A strange thought but one that lingers in my mind far longer than it warrants, and every time I see it on the map I think of the Match of the Day team discussing yet another controversial Premier League decision. I inch the final few

hundred metres of descent, zig zagging across the track in a melodramatic attempt to reduce the incline. I even briefly entertain the idea of purchasing walking sticks, which I have up to this point in my life despised. Evidence, if ever there was any, that the walk is taking its toll.

In the middle of the valley stands the graceless lump of a Carrefour supermarket. Its aesthetics aren't particularly important to me, though. What I care about is its proximity to the path and in this respect it must be heaven-sent. Meandering through the aisles, I channel immense self-restraint to ignore a craving for cereal and fresh milk, but instead succumb to a treat of apple juice and barbecue crisps which set my salivary glands racing from the moment I spot them on the shelves all the way to a bench overlooking the rooves of Bourisp. The crisps don't quite live up to my heady expectations and leave my mouth incredibly dry, which wouldn't be too much of a problem if only I had some water left and I wasn't sitting in the scorching sun. The only shaded spot in the vicinity is occupied by an unyielding couple and after waiting in vain, I descend back into the village in search of a fountain. When I find one I decide to wash my feet there, much to the surprise of passing tourists who gawp at this primitive sight until I spin around and send them scarpering with an aggressive glare in their direction, before heading back up to the viewpoint where I'm planning to sleep.

Sleeping outside in the open with no shelter may seem like an arbitrary decision, but in fact it requires a very logical thought process. You have to assess the suitability of the location regarding shelter and comfort, and then make clear your intentions by setting out the mat and sleeping bag and thereby essentially announcing, without making a huge song and dance of it, that you will be spending the night there. You need to gather your possessions close by in an orderly fashion to reduce the risk of them being stolen. When you get into the sleeping bag you're still very alert for risks. Firstly visible ones, then once they're deemed minimal and you close your eyes, audible ones. Unfortunately the audible ones are more anxiety inducing as your imagination can run wild and you have to open your eyes to disprove them, which obviously doesn't help the process of getting to sleep. For instance, on this occasion in Bourisp, I hear someone nearby say something about goats and I become fearful that the field I'm in will be inundated with them at any minute. Any inaudible conversations are even worse because you assume they concern you being there. Any insects are upgraded in your mind to blood-sucking mosquitos and any dog barks signal an imminent attack. Then after a while you grow too tired to care, to the point that you gradually adopt the other extreme – an attitude of if I die, so be it.

And yet, despite this mindset, by the time I leave Vieille Aure I decide I don't want to stay out in the open again. As much as I hate to admit it, it causes too much angst and, after 19 days on the go it's becoming too much. This means that from now on either I'll have to pay to stay in a gîte or refuge, or I'll have to use the shepherd's cabanes. In either case it will require more planning – each day will finish at a precise point. I will have to sacrifice the freedom of being able to sleep anywhere and the excitement of feeling more immersed in my surroundings, for the comfort and certainty of shelter.

Arrens-Marsous to Bagnères-de-Luchon

*

A phenomenon I'm becoming well acquainted with is the ability to claim temporary ownership of spaces I would normally disregard or scarcely appreciate. Sleeping outdoors is one example, another is using the public toilets. My view of these before setting off was to avoid them at all costs – disgusting places, riddled with germs, only to be used in the utmost emergency. Now, where I'm increasingly dependent on them, I have come to not only be grateful for them, but also to lose my previous fears about contracting some deadly disease merely from breathing the air inside. In fact, I now feel quite comfortable using them, especially as the alternative involves first digging a small hole in the ground and then crouching over awkwardly. Here, in the village of Loudenveille, I wash my upper body in the basin and change my t-shirt so that I can enter a café with marginally more dignity. There has been a fantastic reassurance throughout the whole walk, that each village has not only a public toilet which is on most occasions in very good condition, but also fountains of drinkable water and, invariably, a church which could provide shelter if I was ever in desperate need.

At an ugly roadside café with an interior full of flies and no customers whatsoever, relatively refreshed in my cleanish t-shirt, I manage to remember to order *un café au lait* rather than simply *un café* which in France means an espresso – a mistake I have made on countless occasions and always leaves me feeling like an idiot. The owner and her family sit outside, chatting and laughing, apparently unconcerned about business and happy to leave me there to welcome any potential customers.

Perhaps my lingering in the café is an indication of a sort of apprehensive clairvoyance about that afternoon because, for the first time since leaving Hendaye I really really struggle physically with the last section of the ascent. It's both steep and drawn out, and the cumulative effect of several long days have begun taking their toll. I'm not able to conjure the mental energy required to battle against my weakening body and mind, and I frequently succumb to the desire to stop. At one point it seems as though I'm taking a break every 20 metres or so. As my progress slows I go into a negative spiral, mulling over difficult situations from a previous job many months before that have somehow followed me to this mountain pass in the middle of nowhere and risen to the fore when I least want them. The whingeing voice of the assistant manager, complaining about some tedious nonsense chips away at me with each step. It's a good thing there's no one around because I might otherwise have gained a reputation as the mad walker who is always muttering away to himself.

Eventually I reach the col, shake off the negativity and pass into a strikingly straight and steep valley with distinctive dark marmots and an empty stream bed which suddenly bubbles up from the surface, bringing the whole valley to life in the process. The Cabane d'Esquierry, where I hope to stay, is locked. Other people, no doubt in a similar situation to me, have scribbled messages on the outside including a sarcastic 'thanks for the welcome' and 'yes to bears' – clearly intended as a provocation to the staunchly anti-bear shepherds. So I'm forced to continue down to the gîte in the valley below just as a storm is brewing. From this vantage point I watch as the first claps of thunder spark a mass exodus and throngs of Sunday afternoon visitors are sent packing in a stream of cars kicking up dust and crunching their tyres back to civilisation.

Once I'm settled and showered, during which my big toenail finally comes off completely much to my relief, I scout out somewhere nearby to cook. But just as I do so the rain sets in again and I return dejected with the prospect of eating yet another cold sandwich. To make matters worse, the only agreeable place to sit is at the small desk in my ground-floor room which faces directly out onto the veranda where guests are continually coming and going. So I close the curtain, switch the light off and sit there chewing stale bread and old cheese in the dark in certainly the saddest meal of the trip so far.

4

Bagnères-de-Luchon
to
Gîte d'Esbintz

90.7 kilometres and 6,680 metres ascent

Monday 4th – Saturday 9th July

Halfway

The town of Bagnères-de-Luchon marks roughly the halfway point of the GR10, at 458km into the trip. I'm in good spirits as I arrive shortly after 2pm on Monday 4th July, my 21st day on foot. In the marketplace a bench allows me to take the weight off my feet and a metal barrier acts as an improvised drying rack for some of my clothes which didn't dry last night. Sitting in such a public space I once again become self-conscious about my appearance – in particular my untamed and unsightly beard which has, to my dismay, failed to give me the rugged outdoor look I was hoping for. I deliberate for a long while about whether to shave it off, decide that I should with the idea it would probably make me more approachable and then, when I visit the shops, forget to buy a razor. I take this as a sign to keep the beard until the end and put up with whatever consequences it may bring regarding other people's perceptions of me.

Spending so much time alone leads certain scenarios to play out in my head and I'm reminded of one such scenario in Luchon when I visit the local Lidl. In this imaginary interaction, I arrive at the supermarket just at the moment a live TV report is being filmed on the damage done to local boulangeries by large supermarkets, which undercut their prices and force these family-owned businesses into the ground. The reporter turns just as I'm helping myself to a few pastries, and asks how I feel about my consumption habits given that local bakers are suffering, to which I respond instinctively 'I've walked 450 kilometres to get here so if I want a croissant for 29 cents, I'm gonna fucking have one', and is met by a stunned silence.

In reality, no such confrontation occurs, although I am overwhelmed by the sheer quantity and variety in front of me, and struggle hugely to make even the most basic of decisions. Having eaten almost exclusively tinned or raw ingredients up to this point it's very hard to think about meals. I can only picture items as individual entities – I can't envisage how different foodstuffs might combine. So in a jittery frenzy I start loading up my basket with bakery produce – baguette, bread rolls, pastries, pizza slices – before stopping myself in a moment of reflection, realising that I'm probably going overboard and coyly putting some of it back. Despite this coming to my senses, I still leave with what looks like a whole week's worth of produce just for the one day, and winding my way to the hostel through the narrow streets of the town I bulldoze my way through three brie focaccias and arrive feeling rather queasy.

For a very reasonable price I have a room to myself at the Gîte Lutin and I'm grateful for the privacy and lack of pressure to socialise. It feels like the height of luxury. In fact, it feels excessive. Now that I've become accustomed to a minimalist lifestyle with no infrastructure besides that I was carrying on my back, the idea of my own mattress, duvet, pillow, towel, curtain, basin, mirror, lamp, chair and clothes rail is absurd. I'm not really sure what to do with them. I've paid for them and yet they just stand there gormless and touched. Similarly, there's WiFi but I don't feel connected to anything online – it all seems so distant and irrelevant. Even some stand-up comedy I would normally find hilarious lacks any appeal and it leaves me wondering if the walk is becoming too consuming and sapping away at my character.

The proprietor Russell is, to my surprise, Welsh. When I booked over the phone his voice was interspersed with a deep reverberating 'urmmmm', so I wondered if there was a problem with the signal. In reality, this is just the particularity of his voice and, combined with a nervous laughter and gleaming grin beneath a ginger-tinged beard, it has the effect of making him immediately endearing, though his age is hard to place. He tells me about his pathway to ending up here, having first spent many years working in the Alps. We discuss the difference between the two mountain ranges and between poverty in urban and rural areas – in particular between London and the south of France. At the risk of adopting a problematic patronisation of poverty in Southern France, we agreed that it seemed possible to achieve a better quality of life down here with less money. But I also have to remind myself again of an expression that, until a few days ago, I thought only existed in English: *l'herbe est toujours plus verte dans le jardin de ton voisin* (literally: the grass is always greener in your neighbour's garden). Had I grown up here I might hold the opposite view and envy the public transport and job opportunities we take for granted.

Russell possesses the sort of relaxed and contented personality that epitomises so many of the people who move down here and now call the Pyrenees their home, and nowhere is this clearer than from the way the gîte is run. It has no reception desk nor is there any formal mention of payment on arrival, so I bring it up while we're chatting, expecting him to fetch a sign-in book or a money box, but he just casually says "that'll be €20" and pockets the note I hand over. The catered and self-catered kitchens occupy the same space, the only difference being separate fridges (with only a sign to prevent anyone helping themselves to the catered produce), so that if someone requests a meal Russell prepares it there and then alongside any guests who might be cooking something for themselves.

I'm almost done cooking when Russell suddenly turns to me and proclaims "right, I'm off" and exits via the back door in such an abrupt and unexpected manner that I half expect the police to come bursting through the front door in hot pursuit. I don't imagine the crime rate in the area is especially high but I still expect there to be some sort of locking up procedure. Apparently not. After my meal, I finish the washing up then switch the lights off, noticing that several of the ground floor windows have been left open and I marvel at the trusting nature of the whole place. No doubt this laid-back approach is the best way to operate when dealing with throngs of visitors coming and going on a daily basis.

*

Food
My craving for cereal and milk is answered in ridiculous fashion in the morning when I devour an entire packet of cheap muesli drowned in UHT milk which settles, begins to rumble gently and then expresses a desire to come right back up the way it went down. I refuse to let it happen. You could argue this is an effective way of dealing with the craving because even the idea of cereal feels abhorrent for the following week. It would be nice to say that the experience taught me a lesson in exercising self-restraint but I can't promise I'll act differently in the future given the strength of the cravings that have been arising.

It goes without saying that food has become a central focus of the trip. Before starting the walk I was very aware that I would need to eat a lot to replace the calories I would be burning relentlessly, yet it was hard to fully appreciate until I was in the midst of it. At times I have been able to just eat and eat and eat, and yet over the course of walk I've still been losing a considerable amount of weight. Many walkers, probably the majority, restrict their diet to 'camping food' – boilable, processed stuff. I wanted to avoid this because on the whole it's unhealthy, expensive and often not particularly filling. The upside of this is that I've been able to have a healthier and tastier diet with more fresh food. The downside to carrying things like tins of cassoulet, bread, tomatoes and cheese is that they are neither the lightest nor the most compact items and I have occasionally looked enviously at fellow walkers unpacking compact sachets with lentil stew which magically transforms from a sterile powder to a warm meal.

My shop from Lidl on the morning of Tuesday 5th July was as follows:

- A slab of rectangular cake (Barre Bretonne)
- A packet of croissants
- Two tins of cassoulet
- One tin of beef bourguignon
- A tabouleh salad
- A packet of dates
- A packet of cashew nuts and cranberries
- A packet of salami
- Four apples
- A packet of cherry tomatoes
- A block of goat's cheese
- Three tins of sardines in tomato juice
- A small ready-to-eat pizza
- A croque monsieur

Bear Country

The climb out of Bagnères is a major struggle with all this food on my back which, together with the water, must weigh at least 10kg. The only consolation is the knowledge it will get lighter as I eat! Given this load, I should be prepared for the exertion that's coming my way. After all, the steepness of the incline is plain to see on the map where the path cuts across the contours in alarmingly quick succession. Yet, however well I predict the landscape in front of me, it's always impossible to accurately conjure the feeling of what it will be like until I'm there, panting away, leg muscles straining through the repetitive rhythm. The route I'm dragging myself up must mark the old road which has since been superseded by a smooth and snaking river of tarmac. This takes a longer and shallower route, so that the two routes, old and new, criss-cross over each other as they ascend. At each intersection I experience a small victory, rewarded with a ten metre stretch of relatively flat road, before once again cutting steeply upwards and setting my sights on the next intersection.

The village of Artigue, where both old and new roads finally reunite, looks to have suffered from years of neglect. Unfortunately, instead of giving it an appealing rustic quality, this has left it feeling abandoned and unloved. It is impressive, however, merely by account of its inaccessibility, and this is by no means the first place where I have encountered such a sensation. The idea of constructing those settlements, even those as ramshackle as Artigue, with nothing but muscle power and without modern tools is scarcely imaginable in my mind. Let alone to do so at an altitude of 1200 metres. It's a testament to human determination and resourcefulness that the original constructors could transform the materials on the surrounding land into habitable structures which sheltered them from both blazing summers and harsh winters.

It has been draining enough for me to make the ascent once, so having to make this trip on a regular basis just to bring up basic goods from the valley below would surely drive anyone to want to leave it behind. For the first settlers a simple trip to the nearest village would have been a four hour round trip on foot, and even further to reach the closest market town where they might have been able to sell their excesses and buy other produce. It's a reminder of how much people's lives must have revolved around subsistence and how little free time they would have had. As a result of this inaccessibility, the countryside I have traversed since Hendaye has been scattered with long-abandoned ruins from where the populations had emigrated in search of an easier way of life in towns and cities. Indeed the rural population of much of the Pyrenees has declined significantly – even Bagnères-de-Luchon has a smaller population today than it did in 1836.

Nowadays it's easy to forget because almost every surviving settlement can be accessed by car. But by walking through the landscape I could feel a sense of connection to the inhabitants who had lived and worked here across the previous centuries. I was immersed in it; I saw the evidence of their work and I shared the same network of footpaths they would have used. Sad though any ruins were, there was also something magical about them. That, in their crumbling state, you could somehow gain a more accurate idea of how difficult life might have been. Their inhabitants had worked hard to carve out their lives here, living off the land, and the very nature that sustained them eventually reclaimed their homes in the form of the mosses, brambles and beech saplings that now surged upwards from every crevice.

Above the treeline I climb into a thick mist and all of a sudden I can't see more than five metres in front of me. The lower half of my view is nothing more than a patchwork of yellow, green and brown grass, spiking upwards in monotonous clumps. The upper half is a monotone light grey which, for brief moments, takes on a dazzlingly white colour where the sun attempts to push through, but to no avail. Where lower and upper halves meet they merge into a blurry middle distance of indistinguishable nothingness.

The faintness of the path isn't a problem initially because it follows a ridge, so there's little risk of deviating, but at some point it would be peeling off and skirting the mountainside across to an adjacent ridge. With such poor signage and visibility it's becoming impossible to distinguish the GR10 route from the network of sheep tracks criss-crossing the slopes. To make matters worse several signs in the past week have warned of *Patous* – the Pyrenean Mountain Dogs that guard the livestock against bears – and the dangers of passing between them and their flock. In these conditions it could easily happen. For the first time I have to get the compass out and read the landscape at my feet in order to find my way. I head out in what I believe to be the right direction, hoping to shortly arrive at the adjacent ridge. In the uncertainty the time seems to drag and with each passing minute I become less optimistic and more tempted to retrace my steps.

A very long fifteen minutes passes until I can make out a rocky outcrop which gradually reveals itself to be the jagged edge of the ridge. However, the orientation is all wrong and I stare at my map and compass wondering how on earth they're not matching up, praying that someone would emerge to point me in the right direction. No such luck, I'm well and truly alone up here. Or so I think. I can hear a faint patter of footsteps from a large animal which quickly builds and I realise something is hurtling at some speed towards me, feet thundering against the ground. In a flash I pocket the compass and sprint to a nearby cairn which offers me a small podium of safety. Seconds later, a horse comes flying past then shoots off into the mist, neighing frantically. I can't tell if it's been spooked by my presence or if there's something else up here. I'm in the heart of bear country all alone, not entirely sure if I'm on the right path and feeling horribly vulnerable. But I can't stay perched on the safety of the cairn forever, hopelessly waiting for the mist to clear, so I plod nervously onwards, still confused about my direction, two more horses shooting past in the subsequent moments. Only when I have to change to the next section of map do I realise I have overlooked the whole section covering the ridge which explains the discrepancy of my orientation and reassures me that I'm on the right track. I'm very glad, half an hour later, to turn off the mountain tops and descend beneath the cloud layer towards my bed for the night.

 The Cabane de l'Artigue is the lowest of the cabanes I had slept in so far, at 1350 metres, but still higher than the UK's highest peak, Ben Nevis. By this point I have grown not only accustomed to, but also very keen on the French habit of displaying altitude at every possible opportunity. The main space of the cabane is taken up by a large table, 2.5 by 4 metres, which looks to be mainly used as a sleeping platform. At its back end three pale yellow and well used sponge mattresses are lined up against the wall. In one corner an unstable fireplace has been elevated from the ground using a dry-stone base, and above it a wide concrete chimney draws away the smoke and must be fairly effective as the smell isn't nearly as strong here as it has been in other cabanes. A generous pile of firewood is stacked beside the fireplace and from one of the beams hangs a sign typed in faded Comic Sans font addressed to '*promeneurs, randonneurs, montagnards*' to freely light a fire on the condition that they restock the pile. Signed 'Ludovic, *le vacher*' and dated June 2018. Two clothes lines also have been strung diagonally from the beautiful exposed timber frame of the roof. Into almost every square inch of this the names and dates of visitors have been written or lightly engraved, one as old as 1971. Annotations include 'EAT THE RICH', 'Jack is Back' and, much to my amusement, here in the middle of nowhere so far from civilisation 'Leeds Leeds Leeds'. On a ledge stand several wine bottles, a pair of Lee Cooper jeans, a jar of screws and an emergency foil blanket. And finally a few home comforts – bin bag hanging from a hook, a dustpan and brush, broom shovel, grill, spare sleeping bag and the rather haphazardly assembled bench where I'm sitting.

Once again any hopes of a good night's sleep are dashed – this time by dormice. I try to shut them up by imitating cats and dogs with a series of ridiculous barking and mewing sounds but either they're not intimidated by dogs and cats, or my impression is not convincing enough. After a lot of resistance I haul myself up and use the broom to smack the beams half-heartedly which works for a few minutes but they soon return to scuttling around screeching at me excitedly while playfighting with each other.

*

I wake up to the sound of rain and have to summon a good amount of energy to release myself from the cosiness of the sleeping bag. Before setting off I perch on the bench, gazing out of the opened front door, whose frame is filled completely by the sloping pasture and pine woods that rise above, imagining that at any moment a bear will come walking through the scene.

Monotony

The GR10 by its nature consists of a series of ascents and descents but the Ariège region, in particular the stretch from Bagnères to Seix, is relentless. No sooner do I reach a valley bottom, than the route starts to tackle the next climb. At many times I start to think about the expedition in very methodical terms – let's get up this section and down the other side – and this pattern can become very entrenched in my mind. On the ascents I can be completely consumed by the ridge above. On the descents, by the river beneath.

Before the beautiful village of Melles a woman in a 4x4 offers me a lift and I explain I can't because I want to cover every metre. "Even with this rain?" she replies. "Fine, if you can take me to the top of the valley that would be fantastic". Only joking! I politely turned it down, only to fixate on how easy it would have been to cover the distance had I accepted.

It's a relief to arrive at the Refuge de Jacques Husson after a long climb of 1,700 metres ascent which, towards the end, crosses a boggy plateau covered in cotton-grass that reminds me of the Pennine Way. The refuge is a relative metropolis with the babbling of conversation and a few members of staff flying around in response to requests. A loaf of bread here costs €20 – more than staying the night – apparently due to the 'delivery price', which I guess makes sense given that everything has to be flown in by helicopter. I deliberate briefly then opt for a quarter of a loaf, but it's the sort of bread that is 90% air so I feel a bit ripped off.

The cabane where I intend to stay is just on the other side of a reservoir, but the hillside is once again shrouded in dense cloud so it feels like a bit of an adventure in itself to make the short journey following the dam wall and up the other side. The sleeping quarters on the upper floor are, to my irritation, packed. Apparently there's a large group on a fishing holiday up here, which at first I take to be some sort of sarcastic joke seeing as there can't be the biggest variety of fish in Pyrenean lakes at 2000m altitude, but turns out to be true. Unsurprisingly they're a peculiar bunch and we get off to an awkward start when one of them asks about the book I'm reading, *The Good Immigrant*. I launch into an impassioned speech about the importance of multiculturalism, which goes down like a lead balloon among an audience who give the impression they think Macron is far too left wing.

Upstairs, I just about manage to find a narrow slot in which to lay my mat and sleeping bag, thereby claiming a spot for the night. Here I meet Allan, who I accidentally disturb from a nap, and who transitions from groggy fatigue to consumingly energetic in a matter of seconds. He's living in Paris and we share a niche joke about the difference between Meudon val Fleury and Meudon la Forêt because I can't resist a childlike urge to show off my knowledge of the city. A few hours later, against the backdrop of a mountainside set alight by the setting sun, I join him and a few other guests for some wine out on the terrace, marvelling at the reality of being specifically *there* in that moment. The wine helps me unwind and heightens my sense of appreciation, which has been waning, though it certainly does not help as I stumble tipsily back across the pitch black hillside in the direction of my sleeping bag.

*

Amidst the blur of images of the scenery that swim through my mind throughout the walk, few are more powerful than the temporary clearings in the cloud to reveal glimpses into the landscape. On these occasions the scene is entirely obscured and only by checking the maps do I have any idea what lies in front of me. Then, in a moment, the sudden unveiling reveals the biblical scale of my surroundings which take my breath away time and time again. Descending into the Vallée du Biros, one such moment occurs as the dense cumulus lifts and rays of light cut through, illuminating the features of the valley floor and making them look all the more dramatic given this shifting pattern in the weather. Individual trees take on a golden glow at their canopies that contrast starkly with the dark blue shadows at their bases. Gulleys carve deep lines into the hillsides, interrupted by the various paths and animal tracks that run perpendicular to them. The grassy plains at the valley bottom become fields of fluorescent green and between them snakes a torrent of water that from this distance seems nothing more than a series of threads, glimmering here and then among the milky cascades and stubborn boulders.

    Despite the stunning scenery that surrounds me, it's telling that most of my photographs from the day are of the hut in the evening. Admittedly, this is partly because the view from the spot I choose for lunch is blocked by a pair of truly special walkers who choose to sit about five metres in front of me, not only completely ruining the scenery but also the solitude. Mainly, though, it's down to my attitude. I'm determined to persevere through the 2100 metres of ascent and a similar amount of descent, becoming largely numb to aesthetic appreciation that afternoon, until I stumble down to the Cabane du Trapech du Milieu, feet really bearing the brunt.

*

At 6:30am the sun acts as a natural alarm, rising above the ridge opposite and streaming right into the empty stable that has sheltered me. Unfortunately, I must have spent too much time looking into it because, for the first hour of walking, I have the sensation I can't see properly out of my right eye. It feels as if my field of vision has been severely restricted and I keep slipping on things I haven't seen which makes me nervous on such a steep descent. Quickly, my mood spirals downwards to the point where it's even tricky to fake a smile to the walkers coming the other way just before the Maison du Valier. Once again, without a voice of reason to tell me otherwise it becomes all too easy to reach for the exit, consider cancelling my reservation for the gîte that evening and getting as far away from the Pyrenees as quickly as possible. In the valley car park I remove my backpack and sit on a bench, trying to gather myself.

    Climbing up the next ascent gives me a view back across the valley to the cabane where I had spent the night – to see the inverse view as it were. I can't help but think about the passing of time. Yesterday evening I had looked out knowing that at some point the following morning I would be able to look back, and now here I am in the moment, feeling as if I've made a sudden jump in time and space. It makes me consider the capability of the human body and the impact of repetitive actions. By setting my mind on a target and placing one foot in front of the other I had, in the space of a few hours, traversed a landscape of quite remarkable scale, simply through innate, relentless repetition.

Beyond the struggles of the early morning descent, it is a day of considerable extremes in my mental state, so my arrival at Gîte d'Esbintz – a superbly rustic organic farm – feels all the more of a relief. The place is empty as I arrive but a sign encourages guests to enter and make themselves at home and I need no second invitation. The room I'm staying in is hugely reminiscent of the gîte we visit as a family during the summer, the same smell of finished wood and a similar array of bedding of various colours and patterns. It's as if my body can sense I'm in the same area of the country. After a glorious shower, the first since Bagnères, I am able to fully relax. Snacks and drinks are available with an honesty box that wasn't even locked or tied down. What a fantastically trusting place! I try to savour rather than gulp a beautifully crisp cider, made on the farm, accompanied by some crisps (organic – of course they were) and feel instantly reinvigorated following what had been a tough few days. Within minutes of arriving I have already thought about asking whether I could return to do a working holiday here in the future. On reflection, this is obviously an indication of my desire to think beyond the expedition, to imagine something in the future that doesn't involve walking.

Over the next few hours a trickle of walkers come through, most surprisingly of all Paul and Hugo who I met at Refuge Jacques Husson, and who must have done epic days themselves to catch up. We all eat together that evening alongside our hosts, Mathias and Adeline, in their beautiful house. Conversations flow from the life of a *berger* and how they survive up in the cabanes, to the appropriation of Tomme and Berth cheeses, to the *balisages* used to waymark the GR10 which are apparently organised by each *département* and maintained by a team of volunteers. I slowly drift off into exhaustion and the speech becomes one big blur as my bed beckons.

*

A Day of Rest
I decide to spend another night at the gîte to give myself a rest day, which is hard to settle on given the desire to press on, to move from place to place, that has become second nature. Once I make the decision, though, it feels like the right one. It gives me an opportunity to relax both physically and mentally, to reflect on the expedition so far and to observe what's happening around me.

Yesterday I glimpsed the ridge of mountains that is so familiar to me from our family holidays here, illuminated by the late-afternoon sun. Somehow walking to the Ariège makes it seem like a different region from the one I am used to visiting on holiday; like an alternate world or a parallel universe that is visually identical but feels somehow altered. Coming down here as a child I was filled with excitement and wonder because it was our annual holiday abroad and an opportunity to experience life in another culture and language. With time, that excitement has tempered and yet the Ariège still occupies a particular space in my mind associated with those holidays. Now, having walked here from the Atlantic Ocean, it feels more connected – part of the wider Pyrenees rather than simply a destination. It's a challenge to explain exactly why this should feel so particular but it is something that plays on my mind every time I look at the map and see a recognisable place name. It just goes to show how our mental state and internal processing of information informs the perception of the world around us.

Once again I am impressed by the energy of my hosts. When they had finished tidying up yesterday they can't have gone to bed much before midnight, but still they're up at the crack of dawn. Mathias leaves at 5:30am with the *troupeau* and Adeline sees to breakfast and assembles a packed lunch for one of the walkers at short notice that is delivered with an indignant death stare. Breakfast consists of the best coffee I have had on the trip so far and bread with homemade jams from the farm – quince, fig and plum.

Adeline mentions that I'm not alone in being surprised by the honesty box system and we all agree there are only a few places that it could work; certainly not in any city. It's laughable, for example, to imagine such a system functioning at my previous job in a charity shop in Tottenham, where our donations box raising money for people with disabilities had to be chained down because it was once stolen twice in the space of a week.

Bruno and Cecile, a couple who it seems insulting to refer to as elderly because they are probably only in their sixties and in incredibly good shape, give me an anti-friction cream and advise me to use it every morning after bearing witness to the horrible sight of my swollen reddened toes the evening before. During the first fortnight most of the friction in my boots was occurring at the heels so I approached any ascents, when the pressure would be mostly at the back of the boot, with apprehension. Luckily, my determined treatment technique of continually applying Compeed plasters until the blisters healed – a sort of war of benign attrition – had worked a treat. But recently a second wave of attacks had come my way, this time at the front of my boots, where the toes would rub against each other as much as the actual boot itself, and this required a more systematic approach with my new advice: part anti-friction cream, part carefully placed plasters with a healthy regime of breaks when the boots could be removed to let my feet breathe.

*

Day 35: Sweet sustenance. Food cravings become particularly acute towards the end of the walk.

Day 39: Silhouetted hills as a storm breaks over the Canigou massif. I pause here to watch it develop and eat ratatouille straight from the tin.

Day 42: Olive trees on the final descent into Banyuls-sur-Mer. I was told about this descent by a walker I met at the halfway point, it feels surreal to have reached this stretch.

Day 42: A skinny, hairy, tanned and happy man standing awkwardly at the Mediterranean.

Day 2: Looking east towards the hills ahead, wondering how the path might navigate through the landscape.

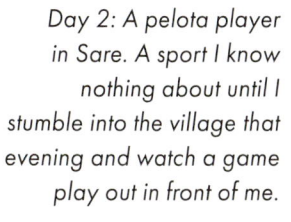

Day 2: A pelota player in Sare. A sport I know nothing about until I stumble into the village that evening and watch a game play out in front of me.

Day 3: The view from the tent at the chapel above Ainhoa. The silhouetted stone cross marks the edge of a burial ground.

Day 4: Le Bastan River before Bidarray, where I meet a small community of walkers including Thomas and Solomon.

Day 4: Sleeping mat set up under a crevice on the ridge below Pic d'Iparla. A cosy spot but also home to a few slugs!

Day 5: The empty town square in St-Étienne-de-Baïgorry with typical Basque architecture.

Day 6: Ascending Munhoa in the morning mist with Thomas and Solomon and struggling to match their pace.

Day 6: Above the clouds on the summit of Munhoa, reluctant to descend back into the world below.

Day 8: Mist in the beech woods on the descent into Logibar.

Day 8: Cooking on the banks of Olhadoko River, Logibar. The following morning it would be in full spate.

Day 13: Clothes drying on a table tennis table in the all-but-empty ski resort of Gourette.

Day 16: Inside the Cabane du Pinet at 2000 metres altitude, m first night after ditching the tent. Hardly a palace, but welcome shelter from the uncertainty of sleeping outdoors.

Day 16: A marmot on alert. These mammals were reintroduced to the Pyrenees in 1948 having become extinct here more than 15,000 years ago.

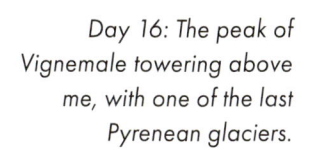

Day 16: The peak of Vignemale towering above me, with one of the last Pyrenean glaciers.

Day 16: The Ruisseau des Oulettes, looking in the direction of Gavarnie down a U-shaped valley.

Day 17: Walking with cows on the descent into Gavarnie, clouds billowing into the valley.

Day 19: Pic de Néouvielle from Lac d'Aumar. Perhaps the most beautiful section of the walk if I had to choose one.

Day 22: The view as I wash my face in a trough outside the village of Sode.

Day 23: Mountainside ablaze during sunset above the Refuge de l'Étang d'Araing.

Day 24: The white and red GR10 waymarks, or balisages. These marks are on a wooden sign, but they are mostly painted directly onto rocks or trees to signal the route, thereby reducing the number of signs needed.

Day 26: The town of Seix in the Ariège region. A small detour in search of supplies during my only rest day of the trip.

Day 28: Lunch and clothes drying on a rock. Removing my sweaty socks and t-shirt when I stop becomes almost a ritual.

Day 28: Sunset from the Cabane de Guzet where I feature on an episode of a podcast made by fellow GR10 walker Allan.

Day 29: Tender feet before a wash in a river. My feet become heavily plastered for the second half of the walk, at one point with a separate plaster on each toe to reduce rubbing.

Day 31: Clothes drying at a washhouse before Siguer. There were many places to wash clothes, but finding the time to dry them proves more challenging.

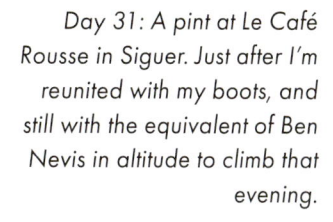

Day 31: A pint at Le Café Rousse in Siguer. Just after I'm reunited with my boots, and still with the equivalent of Ben Nevis in altitude to climb that evening.

Day 31: A cow silhouetted against the ridge above the Cabane du Besset d'en haut.

Day 32: The well-designed Jasse d'Artaran cabin near the Plateau de Beille. It even features a phone charging point powered by a solar panel on the roof.

5

Gîte d'Esbintz
to
Mérens-les-Vals

168.3 kilometres and 10,534 metres ascent

Sunday 10th – Saturday 16th July

Accident Prone

I leave Gîte d'Esbintz in good spirits, glad to be away from a large irritating group with whom I shared the dorm last night. The terrain is fairly easy-going to begin with. Mostly on roads where, after so much time on rough terrain, it feels like I'm floating along. I eat up the first few kilometres with ease, ignoring the increasingly sweaty sensation coming from my boots until I reach a bench by the side of the road, that appears like a God-given sign that I should stop. It's a good thing I take a break then, because for the following few kilometres I am mobbed by horseflies, at which point I have two choices. Either flail around in an attempt to kill them, which is far easier than killing an ordinary fly but takes a while, attracts more horseflies from further afield and gets you considerably worked up; or do nothing, but continue walking briskly which reduces the risk of getting bitten, but does not completely deter them and, as such, also gets you considerably worked up. At first, I try tactic one, stopping to kill nineteen of them in the space of a few minutes, leaving a mass of miniature contorted bodies lying at my feet, but they just keep on coming in a tidal wave of buzzing. So I switch to tactic two with the comfort of knowing at least I am covering ground this way, all the while trying to remind myself that if ever I buy a holiday home in the Pyrenees it should not be in this valley.

Impressively, almost a month in and I haven't had any serious slips or injuries, but that was unlikely to last the whole trip. In my first such incident I step on a rock that gives way and in an attempt to counteract my fall into the small drop below, I sort of throw myself sideways onto the path, grazing my knee, forearm and elbow, and feeling slightly shaky as I continue. Later, in the village of Couflens as I prepare my dinner, the gas canister from which I'm heating the food suddenly bursts into flames from its head and, fearing the whole thing might explode and send shards of shrapnel in all directions, I plunge the canister under the adjacent water tap. Clearly I hadn't screwed it on tightly enough so gas was leaking from between the canister and tube, and the flame had jumped across from the stove. Still distracted by this panicky chain of events, I bump my head on a low-hanging beam above the water basin area only seconds later. My clumsiness continues that evening at the gîte when I slip in the shower and only just catch myself before an accident that surely would have resulted in a sprained ankle at the very least. Then I bang my head on another low-hanging beam on the way out of the shower. Perhaps this is a sign that cumulative fatigue is setting in despite my break at Esbintz or maybe I've been placed under some sort of curse. Either way, I spend the rest of the evening in robotic fashion, doing everything in a slow, mechanical manner and crawl into bed amazed to still be in one piece.

*

A Familiar Face

It's Monday 11th July and looks set to be a glorious day. I'm just stepping out onto the porch when I almost collide headfirst with Allan, who had been camping in the field outside. I'm stunned into silence for a moment as I imagined he had dropped far behind since we parted ways at the Refuge Jacques Husson. He recounts his last week or so which include what sounds like a death-defying late night descent down the sheer east face of Mont Valier. Somehow he's lost none of his vitality, despite being fuelled almost exclusively on a diet of rice and lentils. On the map I show him where I'm intending to stay tonight, which is a very ambitious day's walk of more than 2000 metres ascent. So I feel a sense of patronising surprise as I see him pacing up towards the cabane ten hours later, just as the first orange hues of sunset begin forming in the sky behind him.

From there the sky develops into a breathtakingly rich gradient of colour which provides an appropriately atmospheric backdrop while we produce an episode for his podcast about our experiences of the walk so far.

He's interested in what motivated me to come out here and the revelations I may have had up to this point. It's all very instinctive because I'm answering in the moment (as opposed to a book like this, where I have time to review and revise, and even to overthink) and intensified by a large joint which guides our philosophising. I talk about the walk as a combination of push and pull factors. The push being reactionary – a desire to break free from a lifestyle in which I had been feeling trapped and bored, becoming fearful that I was losing control of my life, caught in a repetitive cycle of work and limited free time. The pull being my inner connection with landscape, a desire for the feel of the morning sun on my skin, a swim in a mountain lake, the sensation of using my body. I ignore any mention of the fact my feet really hurt or that if you see 1000 mountains in a row they lose something of their appeal.

There's another solitary walker at the cabane that evening – a woman called Cécile from Toulouse who is travelling across the area for two weeks on foot and by campervan. She's in a peculiar predicament where she's scared of sleeping outdoors in her tent because of the bears, but also worried that the cabane might be full of germs. Both quite significant drawbacks for someone spending the night in the countryside out here. In the end she opts to sleep outside and either she was eaten by a bear which also consumed her entire tent, or she woke up and left early, because in the morning there is no sign of her.

*

A nightmarish descent down towards Aulus tumbles in a

chaotic, unpleasant and roundabout way into the valley, passing some spectacular waterfalls in the process which I struggle to appreciate. We are overtaken by a lunatic carrying nothing but a unicycle on his back who must have come all the way over from the previous valley this morning because there are no other paths on this hillside. I don't take a photograph of him so it becomes one of those moments where you start to wonder if it really happened. Allan is heading into the Aulus to send a postcard, so we say our goodbyes and he plugs in his earphones before skipping down the path and out of sight. I myself have been carrying a postcard since Arrens, either forgetting to send it each time I pass through a village, or being thwarted by the erratic opening hours of French post offices – for example in St Lizier yesterday which was open only on Tuesdays and Thursdays from 9am until midday. On this occasion the idea of extending the route with a detour just for the sake of sending a postcard is out of the question, so it remains securely tucked away on the subsequent climb.

Numb to the Surroundings

The result of day upon day of mammoth peaks, thundering cascades, calming forests, dramatic rock formations, deep-blue lakes and quaint villages is that I am becoming increasingly numb to the surroundings. It isn't something that has happened overnight, rather a sensation that has steadily built in me until the point where I can reach a pass after hours of sweat-toiling climbing and, where I might previously have looked down into the awaiting valley and marvel at the beauty of nature, I now often feel an unwanted indifference to the landscape. The evidence that my notion of beauty is changing is clear as I descend past the astounding Étang Majeur yet I barely register any noteworthy emotion.

This is not only the longest time I have ever spent in the countryside, but it also on quite another level of immersion. And so, a shame though it is to admit, after four weeks of near-continuous rugged backdrops I am often longing for a village, whilst failing to notice the grandeur of my surroundings or even, on certain occasions, actively despising them and questioning how such lumps of vegetation-covered rock could ever have enticed me here.

Short Temper

At the gîte in Marc I encounter the proprietor – a stern-faced woman who apparently has no evidence of my booking, despite my having phoned yesterday and spoken to someone who sounded exactly like her. I try to keep calm and ask politely whether there are any spaces still available for the night, a question which is met with a look of bewilderment, all the while I resist the temptation to say, 'well there's only you and your husband running this fucking place and I definitely didn't speak to him on the phone'. All the more confusing when it turns out there are in fact multiple beds still available, and if I had the energy and the choice I might have got to my feet right there and marched to another gîte nearby where the act of welcoming guests isn't so alien. But I'm exhausted and this place seems to be the only option for affordable accommodation in the vicinity.

Once my bed for the night is secured my attention turns to other matters, principally food. I'm in a bit of dilemma because I intend to hitch-hike to a town further down the valley the next morning, but realise this will leave me with little time to get back and complete the day's walk. So I can scarcely believe my luck when a trio of walkers ask if I would like any of their food as they've had to end their trip early and return to Paris. Instant coffee, soup, boil in a bag meals, cakes, chocolate and high-calorie bread are all up for grabs, so I opt to take the lot and offer to pay for the items, assuming we could agree a reasonable price given that we were doing each other a favour. I make a rookie error of being overly generous and hand over €20, imagining that they might say in return like 'oh don't worry you're doing us a favour, €10 will be more than enough!', but I'm given a suspicious glance in return as if I was trying to haggle. Once again I'm on the brink of releasing a string of expletives along the lines of 'I'm doing you a fucking favour mate, where's my change?'.

My emotions are calmed by the generosity of Cecile and Bruno, who I met at the Gîte d'Esbintz, who invite me to eat with them and another couple, also doing the GR10. It's an absolute feast compared to what I have been putting up with over the past few weeks – a Cantaloupe melon and Parma ham starter, then an omelette, followed by pasta carbonara and a yoghurt for dessert. I feel slightly overwhelmed by this genuine act of kindness, which was offered freely without pressure of reciprocity at a time when I was struggling. If anything I feel guilty and offer to do the washing up in return which is really the least I could do. Despite the warmth of this interaction I have a deep longing for home, perhaps brought on by this temporary insight into the comforts of everyday life.

*

Familiar Territory

I plug in my earphones on the way out of Marc, trying to block out the deafening silence and, at the same time, to prevent a pattern of thought which has built up imperceptibly by the day but so obviously after one month, whereby my mind can only focus on the end of the day, if not the end of the GR10. The path takes a horribly circuitous route which heads southeast into a valley for seven kilometres, only to climb up the hillside and take a U-turn to leave the same valley along a north-westerly trajectory. It's a test to my stubborn desire to stick to the path, especially as there's a viable shortcut that cuts straight upwards. A long, slow and steady slog later I pull into Goulier, with music still playing, which seems to have done the trick.

Just before the village two young people are gathering wild raspberries by a small clearing in the forest. They're doing the GR10 in the opposite direction, from east to west, and have a marvellous vitality which makes me wonder if in future it would be better to walk with a companion. We share advice about our respective remaining sections so I try to sift through my overwhelming memory bank to select and convey those specific moments that might be most useful. It's funny to think that together we have done the whole path, yet we have the inverse experiences. It makes me question how the experience might be different walking the other way. For one thing the sun would rise behind you, pull up above to your left and then set in front of you. But beyond this every landscape and every village would look quite different. You would admire different views, pause in different places, meet different people and remember ascents and descents inversely.

As the trip has progressed I have been seeing fewer walkers I recognise and I've felt less inclined to socialise. In Goulier, rather than immediately seeking shelter, I prepare and eat dinner on a bench, grateful for the peace and quiet. All this time alone is having an effect on me which acts in a detaching way. After long periods of solitude I often long for an interaction with someone, anyone, to make that human connection, but the interactions I have seem fleeting and isolated. When I spend a long time in the company of the same person or group I will often long for time to myself. It takes me a while to realise what I'm missing are familiar faces, the comfort of a close friendship, a well-established bond. As I walk in to the gîte around 8pm I overhear a murmuring from some fellow walkers *"C'est Mathieu qui arrive"*, and I offer a meek smile before heading to the seclusion of my room.

*

In the valley just outside Siguer an old *lavoir* with a beautiful slate roof of layered concentric tiles, shimmering beneath the midday sun like the scales of a fish, offers the perfect opportunity to wash and dry my clothes, that is long overdue. I make the foolish assumption that there will be a shop in the valley where I can stock up on supplies. Instead, there is something referred to as an *épicerie sèche* and a young waiter at the village bar gives me some incredibly complicated instructions on how to get there, describing it as *"comme un sort de escape room"*. It's essentially a small, locked room with a limited selection of goods and an honesty box to leave payment. To enter I need to phone the owner, receive a code and provide a list of the things I want to buy to make sure it tallies with the amount of money I leave. It's already pissing me off and I haven't even gone inside, so when I see that tomatoes cost 50 cents and it's four euros for a measly, processed loaf of bread I make an instinctive decision to hitch-hike down to the main valley and the nearest supermarket. All of a sudden I feel energised on a new mission and it turns out to be remarkably easy. In a matter of seconds I see a guy getting into his car, so I scurry over and ask if he can give me a lift. Only while we're speeding towards Tarascon-sur-Ariege do I realise that I've left my walking boots at the bar.

Another horrible pang of realisation comes when the driver reminds me it's 14th July – the *Fête Nationale* or France's national day – and all shops are almost certain to be closed. Sure enough the centre of the town is deserted. In a desperate search online, scanning Google Maps, I find an Aldi on the outskirts which is apparently open, though I've learned not to trust online opening hours here, so I try not to get my hopes up. It's a sweaty walk along the main road, feet slipping in my sliders with each step, until I approach the Aldi car park which looks worryingly quiet. I frantically ask a group getting into their car "*C'est ouvert?*" only to be given a gormless stare in return, so I scream "Is it OPEN?" to which I receive a series of nods and anxious smiles as they likely wonder what they've done to be on the receiving end of such aggression.

It isn't long before I'm reunited with my walking boots at the bar, having been given a lift back up to the village, astonishingly by the first car I flagged down. After this successful shopping excursion against all the odds, I can't help but feel just a little smug and reward myself with a blonde beer, sitting down just as a fellow walker comes by and joins me. My intended destination is still 1200 metres ascent away, which I reason to be around three hours climb, and it's an enormous struggle to get going after the drowsiness brought on once again by the beer. I plug into a podcast on Mayan civilisation and plod away until I reach the ridge and rise above the forests in time for golden hour.

As the evening stretches to its end it pulls an inky blue blanket across the valleys that emphasises the undulations in the land, gradually obscuring them as the sun dips to the horizon behind me and then edges below. The multiple ridges I have passed are silhouetted one after another, becoming paler and paler as they disappear into the distance. I look down at my legs and feet, patiently motionless while I scan the land over which they have carried me across, before I turn and continue towards my bed for the night, a small breezeblock and render cabane sheltered on the eastern side of the ridge. Through the window at the end of bed I can see the twin peaks of Pic de St Barthelemy and Pic de Soularac. I remember scaling their summits, over ten years ago now, and staring out at the endless crests and ridges opposite, wondering if I would ever wander along them. It feels special, if slightly surreal, to finally be here.

*

When I hear there's a Nordic village up on the Plateau de Beille it registers in my mind as one of two possibilities: is it a historical site where evidence of our early ancestors had been found – hunting tools, settlements, and other such archaeological gems? Or a sort of mock Lapland-esque experience complete with reindeer, Inuit style tents and pristine countryside? It's somewhat disappointing, then, to find that it's actually a ski resort. The collection of buildings at the northern end of the plateau is one of those places so remote you wonder how humans could have made it into such an eyesore. A spooky dog training place, half built hotels, overgrown car parks, wide tracks bulldozed insensitively through the pine trees and, of course, all manner of horrible ski infrastructure that only looks appealing when buried under a foot of snow. I therefore continue swiftly past, along the ridge that boasts some of the most spectacular scenery of the whole expedition. From this vantage point, I'm able to appreciate it properly, glad to have some temporary respite from the pattern of up and down.

*

6

Mérens-les-Vals
to
Banyuls-sur-Mer

205.5 kilometres and 9,324 metres ascent

Sunday 17th – Monday 25th July

Pyrénées-Orientales

The village of Mérens-les-Vals, which marks roughly three quarters of the way, holds particular significance in my head. If I reached here I would be able to complete the walk. Or so I reasoned over the past few days. On reaching here, though, it's hard to be convinced by this and my attention immediately turns to the next milestone.

There's a steady stream of people climbing up from Mérens, many of them visiting the famous thermal pools here which possess an alluring shade of blue, like tinted glass, but come with the drawback that they stink of sulphur. Among the walkers I come across a competitive guy who asks me when I started in Hendaye, before telling me he started four days later than me. I should have ignored this comment but it got under my skin, so I decide to get revenge by hanging on his tail for the whole ascent, only a metre or two behind, waiting patiently until he eventually wanted a break and offering a smile as I overtake. Childish, maybe, but also highly enjoyable.

The Pyrénées-Orientales region hits me like a slap in the face. From the mountain pass at Coll de Coma d'Anyell it gets markedly drier which has a bigger impact on my mood than I expected. Oh how I took those verdant, towering deciduous trees and their plentiful shade for granted! The soil is a more sandy consistency and the waterways are fewer and sadder. Looking out south-east towards Pic Carlit, the landscape has a moon-like quality. Besides a substantial tarn it's nothing but barren rock as far as the eye can see and it's hard to imagine any life form at all surviving up here. It seems mirage-like, then, when I see two human shaped outlines reclining in the shade of their umbrellas beneath a craggy outcrop, apparently without a care in the world.

Moments later, crossing a stream, I see two trout darting for cover in the shadowy pools up here at 2000 metres above sea level and it reminds me how often I have encountered wildlife living at such high altitudes. On countless occasions I have approached even higher peaks than this thinking how barren and devoid of life the terrain is, only to be confronted with herds of cows, sheep and horses, as well as marmots, lizards, spiders, beetles and, on rarer occasions *chamois* – a wild goat-like animal – skipping across the rocks. While this may be a venture into the unknown for me, and is taking its toll on my body, so many creatures have made this environment their own in spite of these challenges.

I'm considering staying at the Cabane de Rouzet whose interior reflects the brutal landscape outside – two sheets of polythene marking a sleeping area, an old map, a manky clothesline yet little else. Had I arrived in the midst of a fierce storm or imminent dusk my views would undoubtedly have been different but instead, with five more hours of daylight and my determination to seek out friendlier scenery, I stomp off over the next pass. The Lac des Bouillouses at last offers a change in the scenery with its flickering surface, black crests lined up one after another in obedient regularity. Beneath the dam wall, amongst a cluster of tourist infrastructure, an empty hut beckons me inside. As I drift off to sleep, I'm disturbed on three occasions by people who must have been staying at the nearby gîtes and hotels, intrigued to see what was inside this ramshackle hut. They are confronted by a drowsy man barking "bonjour?!" at them, which is code for piss off, and works a treat.

*

Over the past few days I have felt a growing longing for familiarity and as a consequence I have struggled to resist checking my phone at least a few times a day. On the monotonous track down to the unimaginatively named ski village of Pyrénées 2000, I pause in the shade to respond to some messages from home, including some concern that I might be suffering given the recent scorching temperatures in Europe. It's so clear to see the addictive nature of phones after all this time alone. Early on I was able to completely withdraw from any distractions. Checking messages was more of a nuisance that I felt an obligation to do every few days to communicate that I was doing well. As the walk has worn on, this relationship has gradually switched until it has become the inverse. Instead of giving information about my situation *to* the outside world, I'm increasingly wanting to hear news *from* the outside world.

At an out-of-town shopping centre with a gigantic car park, pharmacy, hotels and massive Casino supermarket I spot Kevin, a walker testing the Hexatrek app who I met in the first week. He is the epitome of Frenchness with his beret, coffee and cigarette. I'm not really in a chatty mood but minutes later I'm face to face with him in the supermarket aisle and it's actually a pleasure to exchange stories about the route. My basket is filling to the brim with items while he only seems to be buying a few things. Apparently most of his food is boilable meals which he is already carrying in a backpack so small it makes me think of the TARDIS from Doctor Who. One thing he does recommend is an energy drink called Tiger which looks absolutely lethal to me, so I nod warmly as he mentions it, while totally disregarding the idea in my head.

The gîte in the nearby town of Planès is more of a hotel with prices that reflect this, so I continue to a cabane Kevin has recommended where he will be camping. It's much further than I realise because in planning the route I had cropped a little section of map in order to fit it all onto one page, as a result missing two kilometres and 400 metres of ascent which feel all the more challenging because they are unexpected. Worse still, this gives Kevin another opportunity to show off about the Hexatrek app. Maybe I'm being won over to its benefits. Nevertheless, the idea of immersing yourself in nature only to end up using your phone to navigate seems ridiculous to me.

Halfway to the ridge I stop to eat a pain au raisin I had bought at Casino. I rest it on my knee to admire it, letting my mouth water, building up an appetite. It swirls alluringly inwards, drawing my eye to its centre. The pastry glistens in the light, revealing the fine layers from which it is composed, rippling at the edges where they have browned in the oven. Plump juicy sultanas are nestled within the *crème pâtissière* on the inside of the swirls. As I bite down my teeth sink into the soft springy pastry and tear away pushing the mouthful around into every crevice, trying to savour the taste. All too soon it's gone and sadly I won't be passing another supermarket, or even a shop, for four days. After a brief moment to digest and convince myself cravings are just a state of mind, I'm back on the move.

The cabane recommended to me comes into view following a very long hour and a half along the path. It looks to already have an inhabitant and, sure enough, he returns from stretching his legs shortly afterwards. Philippe is his name – a stocky bloke to put it mildly, with a handlebar moustache that serves to emphasise his already curious expression. We get chatting yet somehow a few minutes later I nearly end up fighting with him when our simple conversation over where I'm going tomorrow escalates in ridiculous fashion.

"I'm going in the direction of the refuge, I can't remember what it's called", I say, to which he mentions the name of a refuge nearby.

"I'm not sure, I can't remember the name at all. I'm following the GR10 in that direction", pointing up the hillside I will be climbing tomorrow. Which is met by a look of confusion.

"Towards Banyuls-sur-Mer". Further confusion.

"EAST!", I exclaim in an attempt to clear any possible doubts that could be remaining. His response is to get out a map and point me in the right direction. 'I know where I'm going you patronising old git, I've navigated 700km from the Atlantic', I want to roar. In any case there are only two options to follow to leave this place, and one of them would involve retracing my footsteps back the way I've come which I would obviously not be doing. Thankfully there's glorious light outside so I go and read my book in peace and let the steam evaporate from my head.

*

It would be a lie to say I dislike the final week, but there are certainly plenty of moments where I'm counting down the days. If there's some magical hypothetical situation in which I could instantly be transported to the end of the walk but with the sense of satisfaction of having done the whole thing, I would struggle immensely to say no to it. I begin thinking about the remaining days in a particular way to get me through. Only one week left, but the last day will be enjoyable and all adrenaline, so really it's only six days. And for two of those I'll be spending the night at a gîte, so it's only four days of walking with uncertain evening accommodation. And by this time tomorrow it'll only be three days. So goes my train of thought.

The Refuge du Ras de la Carançà is ablaze with drum and bass when I enter, which takes me by surprise but it feels so welcome amidst the ongoing nothingness. It's a beautifully simple stone building, with a first floor dressed in dark timber that stands unobtrusively yet proudly in the middle of the valley. Inside, a fabulously organised chalkboard details *'toutes les infos pour manger, boire et dormir'* (all the information about eating, drinking and sleeping) and my eyes are drawn to the *pique nique* of 1830 calories on offer which consists of *'une salade de riz et legumes, fromage du pays, rosette en tranches, petits gâteaux, fruits secs, la pomme du coin, le pain qui va bien'*. The breakdown of coffee prices also makes me chuckle:

Un café: 10 euros

Un café s'il vous plaît: 5 euros

Bonjour, un café s'il vous plaît: 1.50 euros

There is an infectious energy about the three young guardians so I stop for a café au lait, doing my best to order politely, and try to soak up as much atmosphere as I can before using the pedal-operated compost toilet and continuing on my way. I walk briefly with Linse and Felix, two Germans who are joining as part of the Hexatrek, an initiative they supported through crowdfunding. It's only when they put up their shade-providing umbrellas that I realise I spotted them two days earlier lounging near Pic Carlit during the afternoon heat. It looks absurd to be carrying umbrellas on a walking expedition given their size and flimsiness but, as they explain to me, it not only saves them a lot of water because they are kept cool, but also saves them the sweat, irritation and need to apply suncream that comes with such extended sun exposure. On closer inspection I see they are also specially designed so they clip into the backpack and are built more resolutely than your average fold-down umbrella.

Mantet is a titchy characterful place of beige stone clinging to a terraced hillside whose gentle gradients are exposed by the sun as it sets, providing a stunning descent on the approach. I build up a healthy appetite for a drink but find I don't have enough cash left for the gîte, let alone to also get a beer. Thankfully the bar has a card machine, so I realise I can use cash back and settle down with my *ambrée*. Only when I come to pay does the barman, who doubles up as mayor in this village of 34 inhabitants, explain that cash back doesn't exist in France. Only banks have permission to act in this way, he tells me. "Can't you just let it slide on this occasion?", I ask. "Just imagine I'm buying two drinks and then give the money back"... Not possible. So I head back to the terrace and offer to buy some else's drink if they give me the cash. Problem solved. I feel very pleased with myself.

The gîte, a little further up the hill, is slowly falling apart in every respect: cracks in the render, tiles missing, shower door tilted at such an angle it no longer closed, rung missing on the ladder to the top bunk. At least I'm able to joke about it with my roommate for the night – a young guy, Leo, who has a miraculously large frame given he was also doing the GR10 and looks to be surviving principally on packets of air-dried noodles. He started on the same day as me, so it's a miracle we hadn't already crossed paths. I wonder how many other people there are like this, sharing in this journey yet having no evidence of each other's existence. He tells me he's also doing some writing as he goes along (hang on, that's my idea!) but besides that he's not particularly chatty and I get the impression that, like me, he's reached the point where he's tired of talking to strangers, but not to the extent that he's gone completely silent. So we both make a half-hearted effort at conversation, both wishing we had the space to ourselves.

*

In the morning it seems he's waiting for me to leave but I'm doing the same thing, taking painfully long and in the end we leave at pretty much the same time which neither of us want.

Canigou

With its steep sides and proximity to the Mediterranean, the peak of Canigou was, until the 18th century, believed to be the highest in the Pyrenees. In fact, it's a considerable way from holding this title, but the mountain nevertheless possesses a certain aura, not only thanks to its significance in Catalan culture but also its prominence above the surrounding plains and the sheer bulk of the massif to which it belongs. For GR10-goers it offers a short detour from the main path and an opportunity to glimpse, on a clear day, the sacred waters of the Mediterranean that mark their long-awaited destination.

Leaving Mantet, I still have a full day's walking simply to arrive at its base, let alone consider climbing it. Rarely have I been so glad to set off under such a mundane, cloudy sky. The sun has been baking me dry. In the village of Py, I seek out a bench and find one beneath the church with a view over the deep-orange tiled roofs. The echoes of dogs barking bounce sharply upwards from the narrow network of streets, a cock crows, house martins twitter and flash among the eaves, and at each quarter hour the church bells clang which casts my mind back to all the settlements I have passed through. On a rocky outcrop across the other side of valley a copse of trees stands out in vivid yellow.

When I get to the Refuge de Mariailles around 3pm, Kevin, Linse and Felix are already there having a late lunch. I'm not wildly keen on sitting with them, but the idea of acknowledging them and then sitting elsewhere is too awkward to entertain. Unfortunately for them I'm in a foul mood, brought on by the fact my phone isn't charging, and I can't summon the energy to hide my true feelings and be convivial. The phone fell into a stream yesterday so I figure it probably needs to dry out completely but even with this explanation, I wonder if it's a sign to minimise phone use for the remaining section and concentrate on my surroundings. The prospect of even a day without it feels difficult though – no alarm, no checking the time, messages, booking places in advance, sudokus. It could definitely have happened at a worse time, but this is little consolation.

It's at this point that I make the decision to forget about the peak of Canigou and continue along the GR10 path. The extra ascent to its summit isn't particularly large; in fact it's laughably insignificant compared with what I've marched over until now, but my muscles seem to be communicating a message to my brain along the lines of 'don't even think about making this walk even longer'. My knees, calves and thighs have formed a sort of unified state where they trudge along in a miserable unison that contrasts so starkly with their sprightly outlook five weeks ago. I doubt they could perform many functions beyond the three states of uphill, flat ground and downhill to which they have become so accustomed. Meanwhile my upper body has become fixed in a statuesque rigidity, arms hovering at awkward angles to grip onto the dangling front straps of my backpack as I walk. There's a cruel irony that such intense exercise is in many ways having an adverse effect on my physical fitness.

My intended resting place is tucked away on the other side of a rounded rise in the land – a hut balanced on the edge of a steep sided gorge, looking across to the peaks of the Canigou massif. I arrive very early and try to get some shut eye, hoping that I might sleep right the way through and wake up the following day feeling revitalised, but no matter how much I will it to happen, my body refuses to switch off. The only company that evening is very short-lived. Two passers-by who enter tentatively but must be put off when I say "*bienvenue chez moi*" because they make a quick exit. I eat the final freeze-dried meal that I had bought from the walkers in Marc, this one masquerading as a pasta carbonara. Each time I've either added too much boiling water so it's more like a soup, or not enough and the pasta is crunchy. There doesn't appear to be a middle ground. On this occasion it's the latter and I crunch my way through on the front doorstep of the cabane, starting to re-read *The Art of Travel* and attempting to see the world with the level of intrigue that the author, de Botton, seems to.

*

I wake from an uncomfortable dream involving captaining the school football team, where I embody authority and professionalism pre-match but give the ball away in my first touch leading to an immediate opposition goal. It's tricky not to connect this negativity with my mindset on the trip, even though there is no obvious parallel to be drawn.

On the path I make a conscious effort to stop during the walk in an attempt to savour it. I'm aware the time is shooting past and in the past few days I have been very much focused on the finish line. There's a particular part of *The Art of Travel* which really resonates with me, regarding the difficulty of escaping from ourselves. When we travel we are usually seeking an escape in which we imagine we can create a mental clean slate, leaving any negative thoughts at home in our everyday life. We look at pictures of our destination and focus on the beauty of the unspoilt mountains, or the difference in the architecture or the exoticism of the vegetation. We imagine how heavenly it will be for us to be there, so we picture ourselves suddenly transplanted into the image. What we fail to do is acknowledge that the same preoccupations, worries, frustrations and distractions we experience at home will of course follow us because our inner thinking and being does not dramatically change in a new environment. This can prevent us from enjoying what's in front of us in the way we might have anticipated. In light of this I have a moment where I try and appreciate the present view without being distracted by the struggle of the walk and, for a fleeting moment, it is possible to see it differently, as though through a different lens.

My thoughts about presence quickly escalate to the philosophical. I think about the present moment as a full scale from pure enjoyment of the moment at one end to pure dislike of the moment at the other, with a sort of distracted thought occupying the middle ground. What determines my mood at any one point? It must be influenced by so many factors. A combination of walking-related factors and those that might impact me in my life more generally – whether I'm going uphill or downhill, the quality of path, view, how much sleep I've had, what I've had to eat. And so builds a complex web that I mull over with each step that keeps me occupied on the meandering route.

Fatigue is certainly something that has accumulated within me over the course of the five weeks, and this comes with its own knock-on effects. I become clumsier, knocking my trailing leg on protruding rocks or not reading the path well and then placing my feet at the wrong angles. Each time this happens I have a go at myself, so there's an increasing presence of swearing and irritation. However, there's also a lot of relief when I clamber to the top of what I classify as the last high altitude ascent which really seems to be significant. And an hour or so later, when I reach Refuge des Cortalets to find my phone charging again, I'm almost ecstatic.

Just a few metres before the refuge I see Kevin and the Germans reclining beneath a multicoloured parasol and some sixth sense told me they would be here. Kevin shouts jokily across "are you sure you're not following me?" and the whole terrace turns to face me. I feel like some sort of eccentric celebrity for a brief moment, but conscious his comment makes me sound like more of a stalker, so I offer a warm smile in a gentle attempt to diffuse any suspicions of the latter. Once I've settled in, I'm able to witness the reciprocal moment. There's a really special feeling sitting at a refuge and watching the relief spread across the faces as people arrive. A sensation shared by everyone at the refuge – the thoughts of long-awaited shade, food and drink materialising before them.

Insect Invasion
At the Cabane Abri du Pinatelli, on a torn piece of notepad paper held in place with a rock as a paperweight, there's a warning note about *punaise de lit* – bed bugs. Having never encountered these creatures before I assume this to have been written by someone with an excessive fear of insects. What harm could they do? Big mistake. At first I dismiss the sensation of tiny legs crawling over me as merely a placebo. But after ten minutes of scratching I admit the prophecy is being foretold. I attempt to stay strong, optimistically hoping they might just take their fill and leave me alone, but they keep coming back for more. I turn on the light to see two of them scuttling away towards the cracks. Not quite! SMASH – my fist comes down on them and I'm alarmed to see just how much of my blood they contain. I try to forget about it and go back to sleep but two hours later wake up to itches all over and find another. I give up and migrate to the cold concrete floor away from the bunks.

*

In my eagerness to reach Arles-sur-Tech I make the mistake of going too quickly across the 18km and 1300 metres of descent, assuming my feet would be more than tough enough by this point. It's scorching hot with the triple whammy of approaching mid-afternoon, descending in altitude, and walking into a valley with limited shade. Only when I'm nearly in the town does the rubbing in my boots begin to feel sore, and I remove them to see the red patches on both sides of my feet have spread and are weeping a translucent pus. To compound my mood, the accommodation options are woeful so the prospect of staying here looks bleak. I'm in a sleep-deprived and weakening state where I can't think straight, so I edge towards the centre and find a bench, trying to make a logical decision. I consider hitch hiking to a cheaper village where I could spend the night, but in the end I grab a cool drink from Spar and head to the river to cool off, stumbling upon a sign for a gîte d'etape after the next ascent.

Within half an hour I'm a new man. I've had a wash in the river, offloaded my rubbish, booked accommodation for the evening and the following day in places further along the path, taken money out, and found a bar to charge my phone. Spirits on the up! All that remains is to buy food for the following two days' walks and wait for the heat to subside, so I head to the only bar I can find − a bland beige-stuccoed building devoid of both character and clientele. The temperature on the flashing green cross of the adjacent pharmacy reads 37 degrees and I'm assuming that the locals have, quite rightly, decided to shelter indoors, because the place is a ghost town. It reminds me of the neglected settlements of Belvis which we visited once during a family holiday and seemed to be inhabited solely by stray cats and weirdos. I have a Limocc! lemonade (from the Occitane region) with ice. Strangely there is absolutely nothing to eat here, not even crisps or nuts. There's a TV in the corner tuned into an unfamiliar eclectic music channel which plays Ed Sheeran followed by Cyndi Lauper followed by random old French songs, but I am grateful for the background noise to cut through the sterile atmosphere. Studying the maps for the remaining walking the landscape doesn't look particularly pleasant. I envisage stunted vegetation and modern sprawling holiday developments. Hopefully I can be proved wrong!

No sooner do I begin my ascent out of Arles, than the first rumbles of thunder come echoing over from the Canigou massif. I chip away at the path, fearful of arriving at my accommodation too late, and not wanting to stop and acknowledge the towering cumulonimbus developing behind me. When I do eventually pause for a break the sky, which was clear blue just half an hour earlier, has transformed into a looming mass of grey, and the rumblings now seem to shake the earth and cast a pervading uneasiness. I make the decision to watch this dramatic shift in the weather system, eating tinned ratatouille and bread while I do. It's astonishingly powerful to see the impact of such spontaneous change. Much to my relief the brunt of the storm is not passing over this way so I'm able to carry on without the panic that had driven me hurrying up and down some of the mountains in previous weeks.

The change in landscape is once again so marked it's impossible to avoid. Since descending into the Tech valley, in which Arles sits, the mountain vegetation is replaced by a Mediterranean biome. Olive trees, stunted oaks, arbutus, mastic and cork trees, which offer little shade and cling on to life in the dusty soil to which they are anchored. It's beautiful in its own way, and I have no doubt that had I not been walking for 39 days I would have been able to appreciate it fully, but now it seems harsh and parched.

Mas de la Fargassa, my resting place for the evening, is more verdant than the surrounding hillside thanks to the presence of a small river that delivers a sense of vitality. Once I've crossed the quaint footbridge it feels like I've entered some sort of Dutch hippy camp with dozens and dozens of people who somehow all seem to know each other. Kids flying around, adults preparing food in the kitchen, a group gathered round the fire playing the guitar and a few sitting here and there chatting and drinking. A former pitstop for walkers further up at the ridge has closed down in the last few years and they decided to offer accommodation here out of 'a moral compass'. Interestingly, this moral compass isn't any cheaper than any of the other places I have stayed at. I help myself to a beer from the fridge but I'm confused about who to pay until a guy sitting next to me says, "that's fine I'll take the money". I'm too tired to probe into this questionable payment system, or to mingle for that matter and within fifteen minutes I'm tucked up in bed, trying to resist scratching the increasingly itchy bed bug bites.

*

From the hamlet of Montalba which, for days after, occupies my thoughts as a sort of monastic paradise because of its position perched precariously on an outcrop of rock, there is a climb of 800 metres or so up to the Spanish border and a summit with the lofty name Roc de France. I have to repeatedly take breaks and I tell myself this is the absolute final big ascent. 'It'll be a breeze from here, don't you worry' says my internal voice. An unusual internal dialogue develops, trying to convince myself as though I have a split personality, one of which knows more about the remainder of the walk than the other and is in a position to offer reassurance. Undoubtedly this is evidence, if ever there was any, that this was now not only a battle against my weakening body, but also my state of mind, which has just about had enough of mountains.

The sweltering conditions prevent me from going more than a minute without having to wipe the sweat from my brow. At a trickle shortly beneath the summit I stoop to soak everything I possibly can in the gloriously cool water that bubbles up from the ground, scraping aside the mulch of leaves and earth that muddy the source, but it's cruelly small and only half refreshes me. Up on the ridge I feel indebted to the carpeted canopy of beech leaves blocking any sunlight from reaching the forest floor. But the same leaves prevent me looking out, so I'm unable to get a clear view across to the Mediterranean until a path up to a rocky outcrop rises above the treetops. I'm tempted to climb up and look across to the east where I expect the distant waves will be glimmering, drawing me onwards, but at the same time I'm worried about looking. I don't want to climb up, only to be disappointed if I can't see the sea. Or to see the sea and find that it's impossible to enjoy the walk from then on because the finish line has been imprinted on my mind. The idea of my first glimpse of my destination seems so holy and special that it needs to be saved for a moment when I appreciate it fully. When exactly such a moment will arise is unclear to me.

In the end the view down to the Mediterranean almost forces itself upon me in unexpectedly abrupt fashion. It's impossible not to notice it as I turn onto a straight path framed by arching beech branches at the end of which lie the faint yet unmistakeable waters, occupying a soft space between land and sky. I'm caught in a momentary breathlessness, brought to a standstill while a sense of serenity spreads throughout my body. An idea of completeness. Or almost-completeness. Before long another warp in the relationship between land and path snatches the Mediterranean waters out of view and I'm once again trundling along through the woods, steered by the twists and turns at the mercy of topography.

Mistaken Identity

The vegetation changes yet again and for quite a few hours it looks as though I'm in a tropical forest stretching for mile upon mile. Las Illas is built at the heart of this and has something of a murder mystery vibe about it. Like Death in Paradise if it was located inland. Verdant, somewhat smart aesthetic but very few people around, and those I do eventually encounter have the sort of strong and varied personalities that I can only imagine have been scripted by a director who is always hiding just out of the picture. At first the place is, as I might have predicted given the time of day, yet another ghost town. The only signs of life disappear cruelly as I'm approaching when they clamber into their car and speed off.

At what feels like the hub of the metropolis, there's an unclear sign pointing vaguely to the right of the abandoned looking mairie. Then a wooden door with the word gîte spelled across in a bold red modern typeface. It looks very much closed, though the door isn't locked so I tentatively let myself in. No signs of activity in the foyer space. Into the main room and two backpacks are set down beside one of the bunks in the dorm area and there's an immaculately clean kitchen-living area with all chairs resting on the tables, blinds down and curtains drawn. Back outside I go in search of a manager or any other member of the human race but the few doors I do knock at are answered only by barking dogs. Nor does the village seem to have a signal, so there's no possibility of looking up the phone number for the gîte, let alone actually phoning them. Two pairs of boots are lined up neatly outside, presumably belonging to the same people as the backpacks and it suggests they are nearby, probably having changed their shoes after a day of walking and gone in search of a bar.

I'm literally itching for a shower by this point – the bed bug bites have come out all red and pronounced particularly on my arms – so I return inside to take one. 20 minutes later I've dried off, unpacked my things and there's still no sign of anyone. Washing hanging from balconies implies people are living here and, now that I think about it, there are cars in most of the driveways so their owners must be at home. Surely they can't be still on their siesta, it's just gone 6pm. My mind jumps to the worst conclusion. There's been a massacre here shortly before I arrived, the culprit has left the scene and the only sign of life left here is a suspicious, ragged looking foreigner who the police can easily pin the blame on.

All of a sudden an energetic young guy comes bounding in, which really catches me off guard. From what he says I deduce he owns the place, though he doesn't introduce himself as such. I get the impression he's delighted to shock me in this way and I guess this sort of thrill is just about the most exciting thing that ever happens here. He's friendly enough though, telling me about the *Haute Randonnée Pyrénéenne* and showing me a potential shortcut for the next day's walk which used to be the former route of the GR10 (which, guess what, I politely turn down).

The village bar is situated in the hotel, a building which seems excessively large. They have no beers on draught so I opt for a bottle of Leffe which comes served with a wine glass. The only two people around besides the barman tell me in drunken enthusiasm that I should try the beautiful large terrace out the back. Surely a joke given that a large terrace is an unlikely feature in a place with no customers, and if it was so good why wouldn't they be sitting out there? But bizarrely it's true. A beautifully shaded space at the edge of the woods with dozens of tables set, as if throngs of customers would be arriving at any moment, and yet remaining empty on a Saturday evening in the middle of summer. What a mad place. Maybe it really is the set for a film. That or there's some illicit activity keeping this whole place afloat.

*

A combination of heat and bed bug bites make sleep difficult. After twelve hours in bed, though, I wake up feeling less drained than I feared. The morning's route adopts tracks and roads, so it's pretty easy going. I stop worrying about my feet and my attention turns to the bed bugs. My anxiety heightens when I pass the couple from the gîte who tell me what a serious problem they can be and a horror story from her work where a colleague came in crying every day because of an infestation at home. Far from the most comforting thing to say in the situation, but at least I've been warned! I start to feel stupid for not having done my research when I saw the note, guilty about having potentially spread the bugs elsewhere, and anxious about the possibility of bringing them home with me. This dominates my background thought that morning and there isn't much in the way of scenery to distract me, just the same monotonous track and scrubby vegetation.

The impressive 17th century fort at Le Perthus is visible from quite a distance, but the town is obscured in the valley so that I round a corner and almost drop into it. I take back what I said about Gourette. This is hands down the worst place I have passed through on the walk, possibly ever. It's essentially a shopping complex with a ridiculous number of large stores selling mostly alcohol and other goods, leaving small sections of the shop dedicated to everything else. I'm reminded of things I had largely forgotten about. The smell of rubbish, honking of cars, fast food outlets, in-your-face advertising and claustrophobia. As with a lot of border towns it also feels highly suspect, with a sort of loitering that differs from rural types who hang around in French villages, and I suspect a heck of a lot of drugs pass through here.

I decide to tackle the bed bug problem head on, removing everything from my bag to lay it out on the ground in front of the mairie, meticulously inspecting each item before putting it to one side. Apparently the bed bug eggs look like pearl barley, but a couscous spillage from earlier in the trip doesn't help the situation. After trying to remove every visible crumb and bit of dirt, I eventually finding a dead bug, ironically in my hygiene bag, so I chuck the bag itself along with a fluorescent t-shirt, a horrible freebie from a football tournament in Rome, against which it's impossible to distinguish anything light-coloured. Remarkably everything else appears fine and I'm able to re-pack my bag feeling considerably less anxious.

In the midst of this painstaking process a man approaches, toastie in hand, gesturing to me. At first I can't understand what he's implying but then I realise he's mistaken me for a homeless person. It shouldn't be massively surprising given my shaggy beard, worn clothes and grubby backpack, but the moment comes as a bit of a shock. I graciously decline, tempting though it looks. He responds by digging his hand into his pocket and offering me a handful of coins. "Here, take this". "*Non merci, ca va*", I reply. I don't want to create any awkwardness by telling him that I'm not homeless, I've just neglected my personal hygiene for five consecutive weeks, but equally I couldn't bring myself to accept his generosity. The lunch I do eat a few moments later – a ham, camembert and tomato baguette – tastes awful because all the ingredients have been maturing in my backpack for 48 hours and, for the first time on the trip, I don't finish my food. Maybe I should have accepted the toastie after all...

My search for a bar in which to shelter while the heat subsides ends at a place with pink awning and lime green walls. The aesthetic indoors suggests they stopped mid-renovation many months ago and haven't been bothered to tidy away the tools because they could be restarting at any moment. DIY table in the corner on which stand a drill, spanners, other assorted objects, a cube-shaped TV that looks as if it weighs a tonne, and to top it all off a sad looking children's play area in the corner which I get the impression is only used by the owner's children who are currently being shouted at by their grandmother for talking with their mouths full of food. In any case I'm very grateful for the shade and the smile that accompanies my refreshing lime Fanta and pain au chocolat.

The heat is still fairly insufferable when I trudge out of Le Perthus at around 4:30pm but I can't stand the idea of lingering any longer. The winding gravel track crunches underfoot. Water tanks dotted every now and then amidst the scrub indicate the very real threat of wildfires at a time when there are many uncontrolled blazes across Europe, including an apocalyptic-sounding story I hear from a passer-by about Australian news channels reporting on wildfires around London. A sign in the bathroom of the hostel about paying attention to water usage emphasises the dryness of the environment and, having walked among the sources of water for so many days, I have a newfound appreciation for how precious it is and how delicate the relationship is between urban areas and their water sources.

There's no kitchen here in an obvious ploy to force guests to use the restaurant. That's not going to work with me! I have gas and food left so I prepare my own food on the terrace. What could have been quite a nice meal as far as tinned food goes – Spanish lentil stew – is ruined because I mix it with copious quantities of pasta. Digesting this starchy meal brings on a wave of exhaustion and any previous intentions to go for a drink on the terrace, which has an astonishing view back westwards over to Canigou, are scuppered. Lying in bed it is impossible not to think about the following day and that in 24 hours' time I would be finished, yet the more I do so the harder it becomes to fall asleep.

Mérens-les-Vals to Banyuls-sur-Mer

*

The Finale

The final moments of any major feat feel as if they merit a special ending. Weeks before arriving at this point I was wondering whether I would do anything special on the last day like wake up early to watch the sunrise, but in the end I'm too wholly drained for anything like that. Nevertheless, by 7:30am I'm ready to leave, except that I haven't yet paid. At the terrace there is evidence that people have had breakfast very recently – plates with crumbs, empty glasses and mugs that recently held orange juice and coffee – but no one around. Even stranger, some delivery bags have been left on one of the tables, no doubt by someone in a similar situation to me, who got tired of waiting and chose to leave them unattended. The sliding window by the bar is open and around the back there's an open French window and curtain flapping in the wind. I gently clear my throat and hear a rustling behind, the proprietor shortly emerging groggy faced with an expression that combines gratitude that I've been honest enough to pay and irritation at having been disturbed from sleep.

Towering pines sway gently above me as I make the short climb up to the ridge which I would follow for the remainder of the day. In my mind I imagined I would be looking down onto the Mediterranean throughout the day, so it's disappointing to still see mountains rolling away in all directions, albeit far smaller than those I had become accustomed to scaling on a daily basis. As the morning develops it becomes a source of mounting concern that the sea still has not come back into view. I spend a long time wondering if I've somehow oriented myself incorrectly and that it's far off to my left or right, or even the nightmare scenario that I've missed out an entire section of map again and there are in fact multiple days of walking remaining. In desperate situations one's uncertainty can play horrible tricks on the brain.

It is an overused maxim that the journey is far more important than the finale, and yet to experience this first hand in such a vivid manner, with all its physical and mental consequences is something that no words could have prepared me for. Over the course of the expedition, and particularly since that rest day at Gîte d'Esbintz when I sat down to plan my day of arrival, I have had the beach at Banyuls firmly set in my mind. Not necessarily always tethered at the forefront, but nevertheless present, floating about somewhere in the not too distant headspace, ready to surface at any moment of hardship, evening arrival or other such poignant moment. This point, consciously or unconsciously, is the ultimate certainty of the walk – to finish the GR10 I would have to reach it. And the closer I get the stronger its presence grows, and the stronger my anticipation of it is becoming, like a magnetic force, a salmon finding its spawning ground or a bird migrating to breeding territory. At the same time, as the Mediterranean is drawing nearer, so does the certainty that I will make it, so that by the time the magical day arrives it's hard not to think of those last few kilometres as a formality that I have to get out the way.

It's not until I'm remarkably close that the distinguishable hue of the sea appears amidst the haze and I realise I must have been looking at it the whole time, yet unable to make it out. It becomes hard to pace myself, to enjoy the last few hours, with the thought of the lapping waves playing strongly on my mind. Before I know it I'm striding intently down the final descent, skirting the final hillside, dropping beneath the rugged world above into the olive groves and then the vineyards, the sea taking on an ever-deepening shade of blue. And finally, it's there at my feet.

Postamble

The process of writing this book has followed a rather similar trajectory to the walk itself, though to a less intense degree. The idea was founded with enthusiasm and excitement, to the extent that I became impatient to write and found it hard to imaging tiring of the activity. From there, in both cases, the excitement gradually transitioned to present enjoyment. With walking came the pleasure of my sudden independence, and with writing the pleasure of reliving and processing the events of the trip. This, in turn, became partly replaced with less positive emotions of fatigue, frustration and isolation as the novelty wore off, and required strong will to continue, motivated chiefly by the idea of reaching the end. Not to say that the final sections of the trip or book were completely mired in misery – both were often a source of satisfaction, reflection or appreciation – but these became fewer and further between and more connected with the increasing certainty of the destination.

Reaching the end brought with it, predictably, a sense of immense satisfaction and, for the day after finishing, immense relief. With time, as the immediate experience of the walk became more distant, there was a tendency to reconstruct a rosier narrative than the reality, something that can be both a blessing and a curse. Thankfully the practice of documenting, through diary entries and photographs, has helped me maintain a strong image of the events across the 42 days.

I am often asked whether I would attempt the GR10, or a similar long distance walk, again. If I had been posed this question on the descent into Banyuls, or at any point during the last week, I would almost certainly have said no! Looking back now, with the benefit of hindsight it's easy to respond in the opposite extreme and forget the hardship.

In reality, the walk was tougher than I had anticipated. Not necessarily physically, but certainly the mental challenge after so much time alone took a much greater toll on me than I could have imagined. This in turn impacted my entire experience, making the physical element feel more challenging, sapping at my appreciation of the landscape and reducing my desire to socialise along the way. Of course this battle required significant determination to overcome, and the pleasure at finishing would not have been nearly so rewarding had it not involved such a battle. Going through the experience has also taught me a lot about how to manage my mental health so that, when it comes to attempting another long distance path, I am better equipped to deal with the challenges that come my way.

Despite the hurdles, it goes without saying that it produced countless truly astonishing moments. The difficulty I have had to accurately do justice to the beauty of the Pyrenees is a testament to this. Barring a few isolated sections, the GR10 is a near-continuous exhibition of indescribably beautiful countryside which, coupled with

Postamble

some special interactions and the idiosyncrasy of personal experience, created a magical journey. To anyone wondering whether such an experience would be right for them I say: do it!

Glossary

Ambrée – A type of beer, similar to a pale ale.

Auberge/Auberge de Jeunesse – Hostel/Youth hostel. The auberges I came across tended to be simpler, more independent and with more character than hostels in the UK.

Balisages – Waymarks. A particular method of demarcating the Grande Randonnée routes in France using two marks which can be painted straight onto rocks or trees, saving the need for wooden signposts.

Berger – Shepherd.

Cabane – A hut, often used by shepherds when they're staying with livestock at high altitude, but also left open for public use at other times.

Canicule – Heat wave.

Chamois – An animal somewhere between a goat and an antelope that lives at high altitudes in several European mountain ranges.

Cheveux d'ange – Literally angel's hair. Think spaghetti but thinner and shorter.

Cirque – A steep-sided area at the head of the valley formed by glacial erosion. Also referred to as a corrie in English.

Col – Mountain pass.

Glossary

Demi-pension – Half-board (accommodation with dinner and breakfast included).

Département – Administrative area, the equivalent of an English county.

Épicerie sèche – Literally a dry grocery. The one I came across in the village of Siguer was effectively a large store cupboard with almost exclusively packaged goods with long expiry dates, as opposed to a regular épicerie which would commonly sell a wide range of fruit and veg.

Gâteau Basque – A traditional Basque country cake with a flaky, buttery crust and an interior with crème pâtissière and a layer of black cherry jam.

Grand chistera – A version of the Basque sport pelota, played against a wall called a fronton. Each player wears a wicker glove called a xistera with which they catch the ball as it bounces back off the fronton, before launching it back towards the fronton on one continuous motion.

Lavoir – Washhouse. Most of the villages I passed through had one, a remnant from the days when inhabitants would all wash their clothes at one centralised location.

Le vacher – Cowherd.

Mairie – Translates as a town hall, though almost every village has one and the mairie plays a much more active role in French rural life than the English equivalent.

Périphérique – Paris's ring road.

Promeneurs, randonneurs, montagnards – Walkers, ramblers, mountaineers.

Séjour – Living room.

Troupeau – Flock/herd.

Un demi – A half pint. Slightly less to be precise: 250ml.

Un gîte/gîte d'étape – A gîte refers to a type of accommodation, though its form can vary quite widely. Rentable holiday homes, for instance, are often referred to as gîtes. A gîte d'étape, often also abbreviated to gîte, describes a subcategory which is predominantly a simple, low cost overnight resting place. Accommodation is usually in a shared dormitory with mattresses and pillows. Communal dinner and breakfast are almost always available.

Une pinte – A pint. Slightly less to be precise: 500ml.

Vente d'œufs frais – Fresh eggs for sale.

Acknowledgements

It goes without saying that I couldn't have completed the walk or the book without a support network. Firstly, I am eternally grateful to my family who nurtured and encouraged my love of walking. I feel lucky to have had the walking drug instilled in me from a very young age thanks to the countless family holidays we have been on. If not for these adventures I very much doubt I would see anything appealing in covering 900 kilometres on foot! My family have also been very supportive of my unusual and seemingly impulsive life choices, which has allowed me to pursue many fascinating opportunities at a young age, the GR10 being just one. When it came to the proofreading and editing process my parents offered invaluable help, from spotting embarrassing grammatical errors to assisting with the book design, and everything in between.

On this note I am also indebted to my team of friendly editors. Alex Pemberton, Ben Cupples, Chloe Bidgood, Jess Hammond, Issy McFarlane, Kilometres Oatfield, Lucy Staite, Paul Baker and Sahan Perera-Merry. You have eagle eyes and caring hearts! Also to friends more widely who have shown interest in my projects. A smile, a text or a joke can make such a big difference. I am grateful for them all! On a professional level, thank you to the team at Troubador for agreeing to publish this book and for the all the advice they offered along the way. It has helped immensely! Particularly I would like to thank my Production Controller, Holly Porter.

Of course the walk itself was also full of kind souls at every turn, many of whom did not get a mention in the book despite my occasional grumbling. I am lucky to have met so many welcoming hosts at times when I was in need of a friendly face. In particular I want to thank the manager at the auberge in Luz St Sauveur, Mathias and Adéline at Gite d'Esbintz and the Dutch couple at Mas de la Fargassa. Similarly there were many many walkers whose mere presence could be enough to lift my spirits, and who I could connect with instantly thanks to our experiences of the path: Hannah and Lennie, Thomas and Solomon, Billy and Alex, Kevin, Hendrick, Chris, Michael, Richard, Hugo, Allan, Felix and Linse.

Finally, a few more general mentions. To those in the Ariège who ignited my love of the Pyrenees and who have created many magical experiences: a special thank you to Nigel and Ute, and Andrew, Leila and Eden. To those who help preserve the GR10 and enable us to explore the stunning countryside it passes through – unsung heroes whose names are unknown to me but whose work makes the GR10 possible. And to the countless remarkable people I met in Lebanon, where most of this book was written.

ABOUT THE AUTHOR

Matthew Bowmer was borwn in London. He studied Geography at the University of Bristol with a year in Paris at Sciences Po. Since graduating in 2019 he has worked in a variety of roles in Bristol, London and Beirut. He is an avid walker, having completed the Pennine Way, Thames Path, South Downs Way, GR10 and part of the Lebanon Mountain Trail which he hopes to finish in the near future. In September 2023 he will begin studying for a PGCE in primary education at UCL. This is his first book.